PREDATORS AND PARASITOIDS

Advances in Biopesticide Research

Advances in Biopesticide Research highlights the progress, emerging trends and future strategies that aim to utilise phytochemicals, microbial products and natural enemies of pests, such as predators and parasitoids, as a means of managing insect pests in an environmentally benign manner.

Edited by Opender Koul and G.S. Dhaliwal

Volume 1
Phytochemical Biopesticides
edited by Opender Koul and G.S. Dhaliwal

Volume 2
Microbial Biopesticides
edited by Opender Koul and G.S. Dhaliwal

Volume 3
Predators and Parasitoids
edited by Opender Koul and G.S. Dhaliwal

A title of related interest
Insecticides of Natural Origin
by Suth Dev and Opender Koul

The publisher will accept continuation orders which may be cancelled at any time and which provide for automatic billing and shipping of each title in the series upon publication. Please write for details.

PREDATORS AND PARASITOIDS

Edited by

Opender Koul

Insect Biopesticide Research Centre
Jalandhar, India

and

G.S. Dhaliwal

Department of Entomology
Punjab Agricultural University
Ludhiana, India

Taylor & Francis
Taylor & Francis Group

LONDON AND NEW YORK

First published 2003
by Taylor & Francis
11 New Fetter Lane, London EC4P 4EE

Simultaneously published in the USA and Canada
by Taylor & Francis Inc,
29 West 35th Street, New York, NY 10001

Taylor & Francis is an imprint of the Taylor & Francis Group

Typeset in Arial Narrow by Graphicraft Limited, Hong Kong
Printed and bound in Great Britain by
TJ International Ltd, Padstow, Cornwall

British Library Cataloguing in Publication Data
A catalogue record for this book is available from the British Library

Library of Congress Cataloging in Publication Data
A catalog record for this book has been requested

ISBN 0-415-30665-5

CONTENTS

FIGURES

TABLES

Contributors

John E. Banks
Environmental Sciences
Interdisciplinary Arts & Sciences
University of Washington
Tacoma
USA

Renato C. Bautista
United States Pacific Basin Agricultural
  Research Center
US Department of Agriculture,
  Agricultural Research Service
2727 Woodlawn Drive
Honolulu HI 96822
USA

David Bennett
Centre for Environmental Stress and
  Adaptation Research
Department of Genetics
La Trobe University
Bundoora, Victoria 3083
Australia

Randy J. Coleman
Kika de la Garza Subtropical Agricultural
  Research Center
US Department of Agriculture,
  Agricultural Research Service
Weslaco TX 78596
USA

G.S. Dhaliwal
Department of Entomology
Punjab Agricultural University
Ludhiana-141 004
India

Gary W. Elzen
Kika de la Garza Subtropical Agricultural
  Research Center
US Department of Agriculture,
  Agricultural Research Service
Weslaco TX 78596
USA

DeAnn Glenn
Grape and Wine Development
  Corporation,
Kent Town,
South Australia
Australia

Ernest J. Harris
United States Pacific Basin Agricultural
  Research Center
US Department of Agriculture,
  Agricultural Research Service
2727 Woodlawn Drive
Honolulu HI 96822
USA

Ary Hoffmann
Centre for Environmental Stress and
  Adaptation Research
Department of Genetics
La Trobe University
Bundoora
Victoria 3083
Australia

Edgar G. King
USDA, ARS, Mid South Area
Stoneville MS 38776
USA

Opender Koul
Insect Biopesticide Research Centre
30 Parkash Nagar, Model town
Jalandhar-144 003
India

Ingrid Markovic
Center for Biologics Evaluation and
    Research
Food and Drug Administration
Laboratory of Cell Biology, Division of
    Monoclonal Antibodies
Bethesda MD 20892 HFM 558
USA

R.E. Cruttwell McFadyen
Queensland Department of Natural
    Resources and Mines
PO Box 36
Sherwood Qld 4075
Australia

Juan A. Morales-Ramos
Southern Regional Research Center,
    Agricultural Research Service
United States Department of Agriculture
1100 Robert E. Lee Blvd.
New Orleans LA 70124
USA

Dale M. Norris
University of Wisconsin
Madison WI 53706
USA

M. Guadalupe Rojas
Southern Regional Research Center,
    Agricultural Research Service
United States Department of Agriculture
1100 Robert E. Lee Blvd.
New Orleans LA 70124
USA

Linda Thomson
Centre for Environmental Stress and
    Adaptation Research
Department of Genetics
La Trobe University
Bundoora
Victoria 3083
Australia

S.P. Wraight
USDA-ARS, US Plant, Soil, and Nutrition
    Laboratory
Tower Road
Ithaca NY 14853
USA

PREFACE

In a natural ecosystem, plants, their pests, and the natural enemies of these pests have evolved natural interactions to influence the population levels of any one or all of them. However, once it comes to managed agriculture, these naturally evolved mechanisms are ignored while developing plant protection strategies. This has led to the extensive use of pesticide-based pest management, which has affected our ecosystem and environment. It is, therefore, necessary that, in the current millennium, pest management studies should be bio-intensive. One of the methods in recent years that has gained increased attention is the use of predators and parasitoids having a key advantage in their capacity both to kill and to reproduce at the expense of the pest. The equilibrium population size and dynamic behavior of many phytophagous insects are largely determined by their natural enemies and any reduction of these results in insect outbreaks. Therefore, in a strict ecological sense, applied biological control can be considered as a strategy to restore functional bio-diversity in agroecosystems by adding, through classical and augmentative biotechniques, or by enhancing naturally occurring predators and parasitoids through conservation and habitat management. Hyperparasitism is another aspect, which is a highly evolved behavior in insects involving the development of a secondary parasitoid at the expense of a primary parasitoid. This represents a highly evolved fourth trophic level. Although hyperparasitoids have traditionally been considered harmful to the beneficial primary parasitoids of insect pests, obligate hyperparasitoids should be distinguished from facultative, and exotic from indigenous. If the influence is positive and the extreme oscillations of primary parasitoids are dampened, then some insect hyperparasitoids might even be considered beneficial as the emerging patterns suggest that hyperparasitoid community structure seems to follow the same organizational patterns as that of primary parasitoid communities.

There are a number of successful examples on record showing the number and extent of successes obtained from biological control importation projects against insect pests at global level. The computerized data of IOBC (International Organization of Biological Control) shows that out of 4221 importations, 1251 natural enemies were established, 2038 failed, and the fate of another 932 was unknown. This implies a success rate of about 38 percent. Obviously, mass production of parasitoids and predators is a highly technical job beset with numerous problems. It needs years of experience and knowledge to tackle these problems to ensure a continuous production of various colonies. The areas of work that need urgent attention include quantification of natural enemy biodiversity and standardization of conservation methods, perfection of augmentation, refinement and automation in mass production technology, and enhancement of efficiency of natural enemies.

This volume is, therefore, designed to provide the answers and a detailed account of predators and parasitoids vis-à-vis their use as biopesticidal agents. We have attempted to address the role of natural enemies in pest control as an integrated pest management concept and how far *Trichogramma*, the extensively studied natural enemy of insect pests, has been used as a pest management tool. As mentioned above, perfection of augmentation is an area that needs attention; we have tried to discuss the subject using the present concepts. The most important aspects such as ecological assessment of natural enemies, tritrophic level interactions including inducible defenses of plants, and the influence of plant diversity on herbivores and natural enemies are covered in this volume. Three specific chapters concerning biological control research (multiparasitism effects on parasitization, synergism between insect pathogens and entomophagous insects, and use of exotic insects for weed control) address these topics in detail to illustrate various advances in this direction in recent years.

We are pleased to have contributors from various parts of the world who are very well known in their respective fields of biological pest control. We hope that this compilation will provide substantial state-of-the-art information for further research and discussion. We thank all the authors for their efforts and perseverance, as well as the many peer reviewers whose comments and suggestions invariably led to the improvement of chapters.

Opender Koul
G.S. Dhaliwal

Predators and Parasitoids: An Introduction

1

Opender Koul[1] and G.S. Dhaliwal[2]
[1]Insect Biopesticide Research Centre, Jalandhar, India, [2]Department of
Entomology, Punjab Agricultural University, Ludhiana, India

Introduction

The terrestrial communities comprise decomposers plus three
trophic levels: plants, herbivores, and predators and parasitoids.
In support of this model, Hariston *et al.* (1960) argued that
plant sources are generally abundant and underexploited (the
green world hypothesis). Under this model, biological control
agents are found only at the top of the trophic web and are
limited only by the availability of resources. However, predators
and parasitoids are natural enemies that attack various life stages
of insects, resulting in the regulation of herbivore numbers in a
particular ecosystem. This regulation has been termed "biological control" and has been defined by DeBach (1964) as the
action of parasites, predators, or pathogens in maintaining another
organism's population density at a lower average than would
occur in their absence. In the applied sense, it may be defined as
the utilization of natural enemies to reduce the damage caused
by noxious organisms to tolerable levels (DeBach and Rosen,
1991). In a strict ecological sense, "applied biological control"
can be considered as a strategy to restore functional biodiversity
in agroecosystems by adding, through classical and augmentative biocontrol techniques, "missing" entomophagous insects or by
enhancing naturally occurring predators and parasitoids through
conservation and habitat management (Altieri and Nicholls,
1998).

The potential of biological control by predators and parasitoids
has largely remained untapped because it has been underused,
underexploited, underestimated, and often untried and, therefore,
unproven (Hokkanen, 1993). In fact, the use of predators and

parasitoids should be a primary consideration in any pest management program. Biological control is generally the best method of control on the basis of ecological and environmental considerations. Biological control is self-propagating and self-perpetuating. The successes achieved more than half a century ago continue to work to this day. Biological control agents possess the ability to search for their host (pest). It is perhaps for this reason that biological control has been especially successful in case of pests that are difficult to control otherwise. Nearly 66 percent of total successes have been obtained in case of homopterous insects, which are covered by a waxy layer and are not easily killed by contact insecticides. Another 18 percent of successes have been reported in case of Lepidoptera, a majority of which are borers and internal feeders (Dhaliwal and Arora, 2001).

Another aspect of biological control is hyperparasitism. In emphasizing the "bottom-up" effects of hosts on their parasitoids, the first trophic level shows that both inter- and intraspecific plant variations can influence the ecology and behavior of the second trophic level of phytophagous insects, which in turn is one of the major determinants of the third trophic level of insect parasitoids (Craig, 1994). Thus there is a highly evolved fourth trophic level relationship that exists between entomophagous insects, and this is called "hyperparasitism." It refers to the development of a secondary insect parasitoid or hyperparasitoid at the expense of a parasitoid. The hyperparasitoids can have an influence on "top-down" control of terrestrial herbivores by parasitoids (Rosenheim, 1998) and their action may have shaped the evolution of parasitoid foraging strategies (Mills and Gutierrez, 1996). This type of parasitism has been studied for many phytophagous hosts (Hawkins, 1994), but there is limited knowledge of life-history parameters and of foraging behaviors of hyperparasitoids. This knowledge is, however, crucial for understanding of the role of hyperparasitism in natural populations as well as in pest control (Sullivan and Völkl, 1999). The knowledge of hyperparasitoid ecology is of basic importance for the design of biological control programs. The aphid hyperparasitoid community as a model system to demonstrate typical patterns of selected life-history aspects and foraging strategies at the fourth trophic level has recently been dealt with comprehensively (Sullivan and Völkl, 1999).

The "push–pull" approach is another strategy for controlling insect pests. This was described for the first time by Pyke et al. (1987) to control Heliothis sp. on cotton and subsequently followed by Miller and Cowles (1990) for protection of onions from the onion fly. However, in both cases no consideration was given to natural enemies and a chemical deterrent or toxin was used to repel or kill the pest. The latest strategy of the push–pull concept is exploitation of natural enemies through trap and repellent plants. The repellent plants reduce the pest attack and at the same time increase the level of parasitism on protected plants, resulting in significant increase in crop yield. For instance, the two most successful trap crop plants — Napier grass and sudan grass — attracted greater oviposition by stem borers than did cultivated maize. The intercrops giving maximum repellent effect were molases grass and a legume species, silver leaf. Push–pull trials using trap crops and repellent plants reduced stem borer attack and increased the level of parasitism of borers on protected plants (Khan et al., 1997; Khan, 2001).

On the whole, habitat manipulation, the use of behavioral chemicals, biological trait improvement, the use of feeding and oviposition attractants, relay cropping and the establishment of entomophage parks are the methods recommended for maximizing the effectiveness of the natural enemies. Differential performances of parasitoids and predators on cultivars with different morphological characteristics are well known. Similarly, distinct variations in behavioral responses to chemical cues from different sources increase the level of parasitism (Paul and Yadav, 2003). Continuous misuse of chemicals has also led to elimination of potential bioagents. Therefore, it is imperative to find ways to protect them from extinction. An entomophage park is an interesting approach to conserving the biodiversity of natural enemies. In India an entomophage park of 0.2 ha area was established in Gujarat, where 28 species of arthropods were conserved *in situ* that were natural enemies to several species of insect pests (Yadav *et al.*, personal communication, 2001).

Historical perspective

The use of natural enemies for combating phytophagous insects has a long historical background. The Chinese used nests of Pharaoh's ant, *Monomorium pharaonis* (Linnaeus) to combat stored grain pests, while *Oecophylla smaragdina* (Fabricius) has been used to control foliage feeding caterpillars and large boring beetles in citrus trees since ancient times. Not only could these colonies be purchased or moved from wild trees, but placing bamboo runways from one tree to another facilitated the movement of ants between cultivated trees. The earliest record of this practice dates back to 324 BC (Coulson *et al.*, 1982).

During modern times, the mynah bird from India was imported in 1762 for the control of red locust, *Patanga septemfasciata* (Serville) in Mauritius. By 1770 it brought about the successful control of the locust. In the 1770s the practice of creating bamboo runways between citrus trees was developed in Myanmar (Burma) to facilitate the free movement of ants between the trees for the control of the caterpillars (van Emden, 1989).

The first well-planned and successful biological control attempt was undertaken during 1887–88, when the developing citrus industry in California (USA) was seriously threatened by the cottony cushion scale, *Icerya purchasi* Maskell. No chemical treatments known at that time could control the scale. C.V. Riley, a prominent entomologist, suggested that the original home of the scale was Australia or New Zealand and that natural enemies of the scale should be introduced into the USA from these countries. The idea found immediate favor and, in 1888, Albert Koebele was sent to Australia for this purpose. He soon found a small beetle known as Vedalia, *Rodolia cardinalis* (Mulsant), attacking the scale in the Adelaide area and in November 1888 the first shipment of the beetles reached California. The beetles in this and subsequent shipments were liberated on scale-infested trees (DeBach, 1964).

Within a year, a spectacular and highly effective control of the beetle had been obtained throughout the citrus growing areas of the state at a total cost of less than US$200. To this day, the Vedalia continues to provide a completely satisfactory control of this pest in California and the savings it has brought to the citrus industry are incalculable. The resounding success of this venture and its successful extension to other parts of the globe where this pest was a problem, coupled with its permanency, simplicity, and low cost, generated enthusiastic support for biological control as a solution to other agricultural pest problems.

Since then, many successful examples of biological control by predators and parasitoids in different parts of world have been well documented (DeBach and Rosen, 1991; Hoy, 1994). The recent outstanding example is the control of cassava mealybug, *Phenacoccus manihoti* Matile-Ferrero, by a tiny wasp, *Epidinocarsis lopezi* (De Santis) in Africa. The edible roots of cassava are a staple diet in most of sub-Saharan Africa, providing up to half the daily calories of 200 million people. The mealybug was devastating the cassava plant in late 1970s, destroying as much as 80 percent of the crop in some areas and widespread famine was a real possibility.

A search was made in several South American countries to look for *P. manihoti* and its natural enemies. *P. manihoti* and several parasites and predators were found in Paraguay in 1981. One of these, *E. lopezi*, proved to be an immediate success. It was first released in Nigeria in 1981 and in less than one and a half years spread to areas as far away as 170 km from the original release sites. By the end of 1984, *E. lopezi* was found in almost 70 percent of cassava fields spread over an area of more than 200 000 km^2 in southwestern Nigeria. Subsequently, the wasp was sent to several other countries, where it was extensively released from air in addition to numerous releases from the ground. It was established over a total area of more than 750 000 km^2 in 16 African countries by the end of 1986. Over the next 7 years, the cassava mealybug problem was effectively eliminated from 30 countries (Herren and Neuenschwander, 1991; Herren, 1996; Bellotti *et al.*, 1999).

DeBach and Rosen (1991) have summarized the number and extent of successes obtained from biological control importation projects against insect pests at global level. By the year 1988 it was found that natural enemies had been introduced against 416 species of insect pests and permanent control was achieved in 164 species (39.4 percent of pests). Of these, 75 species were completely controlled and another 74 were substantially controlled. However, 15 species showed only a partial control (reduction in pesticide application by nearly 50 percent). These successes were later extended to other countries by introduction of the same enemies and, taking these into consideration, a total of 384 importations were successful by 1988.

Biological characteristics

Biological control theory is still based on a three-trophic-level model of arthropod communities (Rosenheim, 1998). In analyses of higher-order predator and parasitoid prediction, with influences ranging from enhanced to disrupted biological control, terrestrial communities are unable to support more than three trophic levels and nonrandom selection of study systems may have promoted this view. To visualize the efficacy of biological control systems, we must understand the characteristics of these systems and know the factors that impede the top-down regulation of herbivores.

Predators

The insects that prey on other insects and mites occur in many insect orders, the most prominent being Coleoptera, Neuroptera, Hymenoptera, Diptera, Hemiptera, and Odonata. In addition, there are several species of mites and spiders feeding on a wide range of insects and mites. Some predators use biting or chewing mouth parts to devour their prey (e.g., praying mantids, dragonflies, and beetles), whereas others use piercing and sucking mouth parts to feed upon the body fluids of their prey (e.g., assassin bugs, lacewing larvae, and hover fly larvae). The sucking type of feeders often inject a powerful toxin that quickly immobilizes their prey.

Predatory insects feed on all host stages, i.e., egg, larval (or nymphal), pupal, and adult stages. Many species are predaceous in both larval and adult stages, although not necessarily on the same kinds of prey. Others are predaceous only as larvae, whereas the adults may feed on nectar, honeydew, etc., and among these it is often the nonpredaceous adult female that seeks the prey for her larvae by depositing eggs among the prey. This is because their larvae are sometimes incapable of finding the prey on their own. Many predators are agile, ferocious hunters, actively seeking their prey on the ground or on vegetation, as do beetles, lacewing larvae, and mites, or catching it in flight, as do dragonflies or robber flies. Certain hunters have specially adapted seizing organs, such as the barbed forelegs of mantids or the labial mask of aquatic dragonfly nymphs. Other predators may use various traps to catch their prey, of which the webs of spiders and sand pitfalls of ant lions are perhaps the best known (DeBach and Rosen, 1991; Beckage et al., 1993).

Insect predators are often embedded in a complex network of trophic interactions not only with their herbivorous prey but also with each other (Polis, 1991; Rosenheim et al., 1995). For instance, 36 genera of hymenopteran parasitoids attack various stages of predatory green lacewings (see Canard et al., 1984). Lacewings are also exploited by parasitic mites and flies, and by invertebrate and vertebrate predators. However, the influences of natural populations of parasitoids on the ability of predatory arthropods to regulate herbivore populations have not been examined, though the interactions are omnipresent. According to Rosenheim (1998), the biological control literature is replete with comparisons of a predator-addition treatment with a no-release control. Predators are generally chosen for testing because they are thought to be effective regulators of a pest population, and predators are often tested in isolation from other members of the natural enemies. It is known that some elements of the herbivore community increase in response to predator introduction. For instance, the effects of praying mantids on arthropod communities inhabiting old fields and pastures show that mantids suppressed some herbivore populations, had no effect on others, and caused still other herbivore populations to expand (Fagan and Hurd, 1994).

The searching behavior of predators is important in understanding predator–prey interactions in biological control. Adult predators may be monophagous (feeding on one prey species), oligophagous (feeding on several species), or polyphagous (feeding on many species). Polyphagous predators may shift to the most abundant prey species, which can stabilize prey populations in a community. If highly mobile, polyphagous predators may be effective in providing control of pests in disturbed systems; monophagous or oligophagous predators tend to be associated with prey in more stable systems. While many predators are carnivorous both as immatures and adults, some feed on nectar, honeydew, or plant food as well. Eggs of predators are often deposited by females in close proximity to their prey. Predators often use visual and chemical cues to locate plants with prey, and may also use prey pheromones as cues. Predators often concentrate on the more abundant prey species, which may make them less effective in controlling a pest at low densities but could make them particularly effective in suppressing pest outbreaks (Hoy, 1994).

Parasitoids

The current phylogeny suggests that the origin of parasitism is from a precursor that specialized in larval feeding within tunnels in wood and at least partly upon fungi. The ancestral form of parasitism was almost certainly external, and in many still extant taxa the parasitoid egg is laid near rather than on or in the host (Whitfield, 1998). The majority of the parasitoids utilized in biological control of insect pests belong to two orders, namely, Hymenoptera (especially wasps of the superfamilies Chalcidoidea, Ichneumonoidea and Proctotrupoidea), and Diptera (flies, particularly of the family Tachinidae). Most insects parasitic upon other insects are protelean parasites, that is, they are parasitic only in their immature (larval) stages and lead free lives as adults. They usually consume all or most of the host's body and then pupate, either within or external to the host. The adult parasite emerges from the pupa and starts the next generation afresh by actively searching for hosts in which to oviposit. Most adult parasitoids require food such as honeydew, nectar, or pollen and many feed on their hosts' body fluids, whereas others require free water as adults (DeBach and Rosen, 1991; Altieri and Nicholls, 1998).

Parasites may have only one generation (univoltine) to one generation of the host or two or more generations (multivoltine) to one of the host. Life cycles are generally short, ranging from 10 days to 4 weeks or so in summer but correspondingly longer in cold weather. However, some require a year or more, if they have hosts with but a single generation per year. In general, they all have great potential rates of increase. The parasites may be either solitary (when only one larva develops per individual host) or gregarious (when more than one larva normally develops upon a single host). However, this is not always clear-cut, as many species may develop facultatively as solitaries upon a small host and gregariously upon a larger one (DeBach and Rosen, 1991).

Adult parasitoids seek out their hosts in the environment using a variety of cues from the target host and its habitat. Visual, olfactory, and tactile cues are used by parasitoids to find their hosts. When a host is found, the female lays an egg in, on, or near the host. Parasitoid larvae feed on host tissues but may not kill their host until the host has fed and developed to adult stage. Parasitoids pupate in or near the host, and emerging adults mate and disperse to search for hosts or other food sources (Hoy, 1994).

Opender Koul and G.S. Dhaliwal

Hyperparasitoids

Hyperparsitoids are secondary insect parasitoids that develop at the expense of a primary parasitoid, thereby representing a highly evolved fourth trophic level. The primary parasitoid attacks an insect host that is usually phytophagous but which could also be a predator or a scavenger. Hence, an insect hyperparasitoid attacks another insect that is or was developing in or on another insect host, and this sometimes impacts on biological control of a pest insect (Sullivan and Völkl, 1999). The structure of hyperparasitoid communities has been studied for a spectrum of phytophagous insects (Hawkins, 1994), and the degree of resource utilization in the field is known for most systems of economic importance. There are two multitrophic systems with a well-studied hyperparasitoid ecology: (i) the heteronomous hyperparasitoids of the genus *Encarsia* attacking scale insects and whiteflies (Godfray, 1994), and (ii) the aphid–parasitoid–hyperparasitoid community (Völkl, 1997). In heteronomous hyperparasitoids or adelphoparasitoids, only males develop as hyperparasitoids, attacking females of their own or another parasitoid species (Williams, 1991). Aphid hyperparasitoids generally develop as obligate hyperparasitoids (Sullivan, 1987). Comprehensive details of this phenomenon have been published recently and are therefore not taken up in this volume (Sullivan, 1987; Sullivan and Völkl, 1999). However, it is important to mention here that conservative policy in biological control programs is urged in order to exclude exotic facultative hyperparasitoids, so that they do not interfere drastically with biological control by exotic primary parasitoids.

Biological control strategies

Introduction/importation

The success of Vedalia beetle for the control of cottony cushion scale encouraged attempts at biological control by use of imported natural enemies. Any pest (native or introduced) may be a suitable host for natural enemies obtained from the same or related hosts in different parts of the world. A computerized record of natural enemy imports maintained by the International Organization of Biological Control (IOBC) has revealed that up to the end of 1988, 4221 importations were made, of which 1251 natural enemies were established, 2038 failed, and the fate of another 932 is still not known. Thus 38 percent of the attempts with known outcome have been successful in establishing the natural enemies in an exotic environment (Greathead, 1990).

About 40 percent of the introduced natural enemies have become established in the countries of introduction and provided partial to complete control of important insect pests around the world. An analysis of case histories of failures of 119 introduced natural enemies released 148 times against 84 species of insect pests (Stiling, 1993) revealed that in 23.5 percent of cases the cause was climate related — very hot summers, extremely cold winters, heavy rainfall, and so on. In another 20 percent of cases, existing native parasitoids attacked the released predators or parasitoids, rendering them ineffective. Lack of alternative hosts when pest numbers were low contributed to a failure rate of 17 percent, while nonacceptability of the particular pest strain by the introduced strain of the natural enemy was responsible for a failure rate of 11 percent. The physical unacceptability of host (prey) from the natural enemy resulted in a 9 percent failure rate. Minor causes such as lack of adult food (nectar, etc.), low reproductive rate of the released natural enemy, emigration from release sites, and so on, prevented establishment in 7 percent of cases.

Augmentation

Augmentative biocontrol is focused on enhancing the numbers and/or activity of natural enemies in agroecosystems. This strategy involves mass multiplication and periodic release of natural enemies so that they may multiply during the growing season. There has been considerable progress in utilizing molecular techniques for more precise identification of strains and species of natural enemies. More recently, progress has been made in utilizing molecular tools to distinguish natural variability between and within a species. Focused surveys in contrasting ecologies are used in discovering the native diversity using geographical information systems. All these aspects will help in release programs, which usually differ in introduction patterns in that they have to be repeated periodically and result only in temporary suppression of the pest and are not expected to become a permanent part of the ecosystem. The periodic releases may be either inoculative or inundative.

- Inoculative releases may be made as infrequently as once a year to reestablish a species of natural enemy that is periodically killed out in an area by unfavorable conditions during part of the year but operates very effectively the rest of the year. In this case the control is expected from the progeny and subsequent generations and not from the release itself.
- Inundative releases involve mass culture and release of natural enemies to suppress the pest population directly as in the case of conventional insecticides. These are most economic against pests that have only one generation or at the most a few discrete generations every year.

Massive releases have been attempted in several programs involving various types of natural enemies. In Malaysia, augmentative releases of *Trichogrammatoidea nana* Zehntner at 39 500–44 500 ha^{-1} resulted in reduction of rice stem borer damage by 51–82 percent (Khoo, 1990). Releases of *Trichogramma chilonis* Ishii and *Epiricania melanoleuca* (Fletchei) have been made for the control of sugarcane borers and pyrilla, respectively, in Pakistan (Mohyuddin, 1994). Similarly, mass releases of several parasitoids and predators have been attempted, with varying degrees of success, against cotton boll worm, rice stem borer, sugarcane shoot borer, sugarcane top borer, sugarcane pyrilla, San Jose scale, citrus mealybug, diamondback moth, and tomato fruit borer in different parts of India (Singh *et al.*, 2001; Singh, 2003).

Opender Koul and G.S. Dhaliwal

Several *Trichogramma* species have been put to commercial use in different countries of the world (Hassan, 1993; Smith, 1996). In Europe, the greatest use of parasitoids and predators has been for the control of insect pests in greenhouses (Hoy, 1994). Presently more than 20 species of biological control agents are sold commercially for use in cucumber, tomato, sweet pepper, eggplant, bean, melon, strawberry, and ornamentals. Biological control is applied on an estimated 6000 ha annually. More than 30 companies mass-produce these natural enemies, and the costs and efficacy are comparable with those of conventional insecticides (Ravensberg, 1994).

Entomophagous insects may be improved genetically for climate tolerance, sex ratio, host-finding ability, host preference, increased host range, increased pesticide resistance, and so on. *Aphytis lingnanensis* Compere is an important parasitoid of California red scale, *Aonidiella aurantii* (Maskell), which damages citrus. A strain possessing enhanced temperature tolerance was developed through artificial selection for over 100 generations. A pesticide-resistant strain of phytoseiid mite, *Metaseiulus occidentalis* (Nesbitt) (an important predator of spider mites in deciduous orchards in the USA), called COS strain (carbaryl-OP–sulfur-OP resistant strain) was selected, commercially mass reared, and released in Californian almond orchards. It has been estimated that this strain has resulted in an annual saving of about US$ 21 256 000 to the almond growers alone. Subsequently, the strain was also used on other orchard crops (Hoy, 1990). An endosulfan-tolerant strain of *Trichogramma chilonis* Ishii has been developed in India and is now being marketed under the trade name of "Endogram." This strain has been further developed for multiple tolerance to monocrotophos and fenvalerate. A monocrotophos-tolerant strain of *Chrysoperla carnea* (Stephens) has been developed. These pesticide-tolerant strains are likely to play a prominent role in cotton and other crop ecosystems (Singh, 2001, 2003).

When deploying a target group of biocontrol agents such as entomophages in crop systems, it is important to understand their interactions at different trophic levels. The nature of the interaction between coccinellids and spiders, such as intra-guild predation or competition, caused them to reduce the aphid density to a lesser extent when occurring together than did the coccinellids alone (Yasuda and Kimura, 2001). Such prediction for the overall impact of augmentative biocontrol is very important, along with the parasitoid–host relationship in the target system. Augmentative bio-control agents often express differential efficacy depending upon their interaction on different host plants. For instance, the role of the host plant on parasitism differences in the braconid *Cotesia kariyani*, a parasitoid on *Mythimna separata* (Walker), has been well documented. Similarly, the compatibility/suitability of host plants in favoring the activity of parasitoids plays an important role in parasitization.

Conservation

This approach emphasizes the management of agroecosystems so as to provide a general environment conducive to the conservation and enhancement of the natural enemy complex (Altieri and Nicholls, 1998). Important conservation measures (Hokkanen, 1993) include the following:

- Use of selective insecticides that do not kill natural enemies
- Avoidance of cultural practices that are harmful
- Use of practices that favor survival and multiplication of natural enemies
- Cultivation of varieties that favor colonization of natural enemies
- Provision of alternative hosts for natural enemies
- Provision of refugia and food such as pollen and nectar for adult stages of natural enemies

Plant diversification of agroecosystems can result in increased environmental opportunities for natural enemies and subsequently improved biological pest control. The effects of various vegetation designs in the form of polycultures, weed-diversified crop

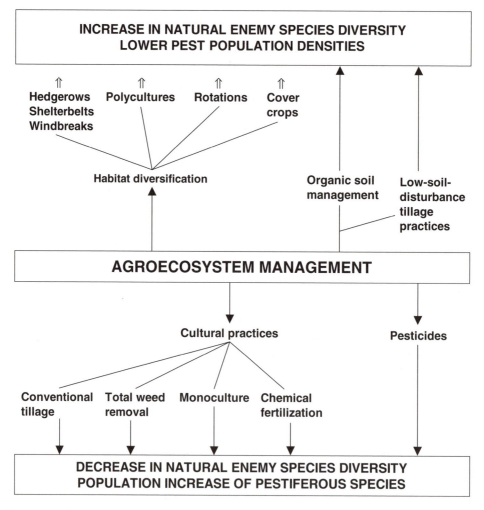

Figure 1.1 Effect of management practices on biodiversity of natural enemies and abundance of insect pests (After Altieri and Nicholls, 1998).

Opender Koul and G.S. Dhaliwal

systems, cover crops, and mulches on pest populations and associated natural enemies have been studied. The factors involved in pest regulation in diversified agroecosystems include increased parasitoid/predator populations, available alternative prey/hosts for natural enemies, decreased colonization and reproduction of pests, feeding inhibition or chemical repellency from nonhost plants, prevention of movement and emigration, and optimum synchrony between pests and natural enemies (Altieri, 1994). There are many agricultural practices and designs that have the potential to enhance functional biodiversity and others that negatively affect it (Figure 1.1). For example, in polycultural agroecosystems there is an increased abundance of arthropod predators and parasitoids owing to enhanced availability of alternative prey, nectar sources, and suitable microhabitats. Such management practices that enhance and/or regenerate the kind of biodiversity that can subsidize the sustainability of agroecosystems through improved biological pest control should be promoted (Altieri and Nicholls, 1998).

Tritrophic interactions

The relationship between host plants and insect pests and natural enemies is based on two hypotheses. The first states that vegetationally diverse habitats support greater diversity of prey and thus have more stable populations of enemies; the second hypothesis points to the relative attractiveness of a habitat to a particular arthropod, based on concentration of host plants or prey species. Plant-produced chemicals play an important role in the host/prey selection process for both parasitoids and predators. Numerous examples are available of natural enemies attacking hosts or prey in one microhabitat but doing so at much lower rates or not at all in others (Nordlund *et al.*, 1988). Indeed, some plants receive very effective protection from natural enemies while others with the same herbivores are relatively unprotected. For example, Nordlund *et al.* (1984) reported that parasitization of *Helicoverpa zea* (Boddie) eggs by *Trichogramma pretiosum* Riley females was relatively high in plots of tomato but almost nonexistent in adjacent plots of corn. Subsequent studies revealed that *T. pretiosum* were responding to plant chemical stimuli and that the presence of volatile stimuli from tomato plants in corn fields resulted in increased rates of parasitization by *T. pretiosum* (Whitman and Nordlund, 1994).

Most plants emit relatively few volatiles until their tissues are damaged and it is the damage that causes the release of these stimuli. When a chewing herbivore such as a caterpillar bites into a leaf, the contents of plant cells and vessels are directly exposed to the air, where rapid volatilization occurs. Not only are existing plant chemicals released, but also entirely new volatile substances can be produced when plant enzymes and protoplasm come into contact with oxygen. Hence, with every caterpillar bite, mini clouds of plant volatiles are released. These volatile plumes drift downward, where they stimulate searching carnivores. They not only alert natural enemies of the nearby presence of feeding herbivores; they also serve as chemical road maps for entomophagous arthropods that follow the scent upwind by chemoanemotactic orientation.

Indeed, many studies have shown that damaged plants are much more attractive to natural enemies than are undamaged plants (Whitman and Nordlund, 1994).

Natural enemies respond not only to chemicals released from plants but also to chemicals produced or released by herbivores themselves. Some of these prey or host kairomones are novel products synthesized by the host, whereas others are plant chemicals that have simply been recycled by the herbivore. In this way, plants influence the volatile emissions and tastes of herbivores, which, in turn, influence the interactions between herbivores and their natural enemies. Volatile compounds emanating from the scales of *Helicoverpa armigera* (Hubner) and *Corcyra cephalonica* (Stainton) were identified as hexatriacontane, decosane, and monocosane, which increased the activity of *Trichogramma chilonis* Ishii, an important egg parasitoid of Lepidoptera. Similarly, kairomones from *H. armigera* eggs influence the egg parasitoid *T. chilonis* and predators such as *Binckochrysa scelertes* (Banks) (Ananthakrishnan, 1993). These chemicals could be applied to the crops in order to attract the natural enemies, and thereby increase the likelihood of encounter with their pest hosts and consequently improve the rate of parasitism/predation. The potential for manipulation of tritrophic interactions to enhance parasitism by application of low doses of toxicants also needs to be explored (Wright and Verkerk, 1995).

Interestingly, these interactions do not necessarily always remain tritrophic. In an ecosystem, these can become polytrophic in nature. Many beneficial insects during their adult stages require a plant food source such as pollen or nectar in order to increase their lifespan and fecundity (Jervis *et al.*, 1993). This establishes a direct trophic connection between plants and beneficial insects, which makes the natural enemy more effective in controlling pest species. This is a polytrophic interaction as the arthropod feeds at more than one trophic level. A classic example of this is increased longevity and higher levels of host parasitism obtained by the braconid wasp *Apanteles* obtaining nectar from weeds surrounding the crop. Using a rubidium-marking technique, it has been shown that a substantial portion of beneficial insects feed on weeds adjacent to an arable crop field (Long *et al.*, 1998).

Outlook

Predators and parasitoids can provide long-term regulation of pest species provided proper management practices are followed to make the environment conducive to furthering their abundance and efficiency in target agroecosystems. Biological control can potentially become a self-perpetuating strategy, providing economic control with the least environmental hazards. However, much work needs to be done to optimize the utilization of predators and parasitoids in integrated pest management (IPM). There is an urgent need to establish a network of large-scale multiplication units so that the natural enemies are made available to the farmers. Heat- and cold-tolerant strains have to be selected/developed in the cases of a number of natural enemies. The environmental implications of releases of these organisms must be studied, especially in cases of introductions and genetically engineered organisms. Since the performance of most biocontrol agents is known to be affected by physical or chemical attributes of the target crops, it is necessary to identify species or strains adapted to major crop systems. It is also necessary to assess the compatibility of natural enemies with the pesticides commonly used against target crops. This will offer improved scope for integration of natural enemies.

In-depth studies are required on the bioenergetics of herbivores, parasitoids, and carnivorous insects. The chemical ecology of the host plant–insect pest–natural enemy interactions needs to be studied to identify the factors favoring colonization by the natural enemies. The use of infochemicals to enhance the behavior of biocontrol agents in agroecosystems holds promise in future IPM. The presence of volatile chemicals promoting the behavior of natural enemies and the fact that the same chemicals affect the growth and survival of herbivores indicate that manipulation of volatile profiles contributes significantly in biocontrol strategies. Some of these issues are discussed in the following chapters.

References

Altieri, M.A. (1994) *Biodiversity and Pest Management in Agroecosystems*, Haworth Press, New York, 185 pp.

Altieri, M.A. and Nicholls, C.I. (1998) Biological control in agro-ecosystems through management of entomophagous insects. In G.S. Dhaliwal and E.A. Heinrichs (eds.), *Critical Issues in Insect Pest Management*, Commonwealth Publishers, New Delhi, pp. 67–86.

Ananthakrishnan, T.N. (1993) Changing dimensions in the chemical ecology of phytophagous insects: role of infochemicals on the behavioural diversity. In T.N. Ananthakrishnan and A. Raman (eds.), *Chemical Ecology of Phytophagous Insects*, Oxford & IBH Publishing, New Delhi, pp. 1–20.

Beckage, N.E., Thompson, S.N. and Frederici, B.A. (eds.) (1993) *Parasites and Pathogens of Insects*, vol. 1, *Parasites*, Academic Press, London.

Bellotti, A.C., Smith, L. and Lapointe, S.L. (1999) Recent advances in cassava pest management. *Annu. Rev. Entomol.*, **44**, 343–370.

Canard, M., Semeria, Y. and New, T.R. (1984) *Biology of Chrysopidae*, Junk Publishers, The Hague.

Coulson, J.R., Klassen, W., Cook, R.J., King, E.J., Chiang, H.C., Hagan, K.S. and Yendol, W.G. (1982) Notes on biological control of pests in China. In *Biological Control of Pests in China*, USDA, Washington, DC, pp. 1–192.

Craig, T.P. (1994) Effects of intraspecific plant variation on parasitoid communities. In B. Hawkins and W. Sheehan (eds.), *Parasitoid Community Ecology*, Oxford University Press, Oxford, UK, pp. 205–227.

DeBach, P. (ed.) (1964) *Biological Control of Insect Pests and Weeds*, Chapman and Hall, London.

DeBach, P. and Rosen, D. (1991) *Biological Control by Natural Enemies*, Cambridge University Press, Cambridge, UK.

Dhaliwal, G.S. and Arora, R. (2001) *Integrated Pest Management: Concepts and Approaches*, Kalyani Publishers, New Delhi.

Fagan, W.F. and Hurd, L.E. (1994) Hatch density variation of a generalist arthropod predator: population consequences and community impact. *Ecology*, **75**, 2022–2032.

Godfray, H.C.J. (1994) *Parasitoids: Behavioral and Evolutionary Ecology*, Princeton University Press, Princeton, NJ.

Greathead, D.J. (1990) Prospects for the use of natural enemies in combination with pesticides. In Jan Bay-Peterson (ed.), *The Use of Natural Enemies to Control Agricultural Pests*, Food and Fertilizer Technology Centre for the Asia and Pacific Region, Tapei, Taiwan, pp. 1–7.

Hariston, N.G., Smith, F.E. and Slobodkin, L.B. (1960) Community structure, population control and competition. *Am. Nat.*, **94**, 421–425.

Hassan, S.A. (1993) The mass rearing and utilization of *Trichogramma* to control lepidopterous pests: achievements and outlook. *Pestic. Sci.*, **37**, 387–391.

Hawkins, B. (1994) *Pattern and Process in Host–Parasitoid Interactions*, Cambridge University Press, Cambridge, UK.

Herren, H.R. (1996) Cassava and cowpea in Africa. In G.J. Persley (ed.), *Biotechnology and Integrated Pest Management*, CAB International, Wallingford, UK, pp. 136–149.

Herren, H.R. and Neuenschwander, P. (1991) Biological control of cassava pests. *Annu. Rev. Entomol.*, **36**, 257–283.

Hokkanen, H.M.T. (1993) New approaches in biological control. In D. Pimentel (ed.), *CRC Handbook of Pest Management*, CBS Publishers and Distributors, New Delhi, pp. 185–198.

Hoy, M.A. (1990) Genetic improvement of parasites and predators. In Jan Bay-Peterson (ed.), *The Use of Natural Enemies to Control Agricultural Pests*, Food and Fertilizer Technology Centre for the Asian and Pacific Region, Taipei, Taiwan, pp. 223–242.

Hoy, M.A. (1994) Parasitoids and predators in management of arthropod pests. In R.L. Metcalf and W.H. Luckmann (eds.), *Introduction to Insect Pest Management*, Wiley, New York, pp. 129–198.

Jervis, M.A., Kidd, N.A.C., Fitton, M.G., Huddleton, T. and Dawak, H.A. (1993) Flower visiting by hymenopterous parasitoids. *J. Nat. Hist.*, **27**, 67–105.

Khan, Z.R. (2001) Biointensive management of cereal stemborers in Africa. Exploiting chemical ecology and natural enemies in a "push–pull" strategy. In S.P. Singh, B.S. Bhumannavar, J. Poorani and D. Singh (eds.), *Biological Control*, Project Directorate of Biological Control, Bangalore and Punjab Agricultural University, Ludhiana, India, pp. 35–43.

Khan, Z.R., Ampong-Nyarko, K., Chiliswa, P., Hassanali, A., Kimani, S., Lwande, W., Overholt, W.A., Pickett, J.A., Smart, L.E., Wadhams, L.J. and Woodcock, C.M. (1997) Intercropping increases parasitism of pests. *Nature*, **388**, 631–632.

Khoo, S.G. (1990) Use of natural enemies to control agricultural pests in Malaysia. In Jan Bay-Peterson (ed.), *The Use of Natural Enemies to Control Agricultural Pests*, Food and Fertilizer Technology Centre for the Asia and Pacific Region, Taipei, Taiwan, pp. 30–39.

Long, R.F., Corbett, A., Lamb, C., Reberg-horton, C., Chandler, J. and Stimman, M. (1998) Beneficial insects move from flowering plants to nearby crops. *Calif. Agric.*, **52**, 23–26.

Miller, J.R. and Cowles, R.S. (1990) Stimulo-deterrent diversion: a concept and its possible implication to onion maggot control. *J. Chem. Ecol.*, **16**, 519–525.

Mills, N.J. and Gutierrez, A.P. (1996) Prospective modeling in biological control: an analysis of the dynamics of heteronomous hyperparasitism in a cotton whitefly–parasitoid system. *J. Appl. Ecol.*, **33**, 1379–1394.

Mohyuddin, A.I. (1994) Commercial sugarcane. *IPM Working Dev.*, **2**, 4–5.

Nordlund, D.A., Lewis, W.J. and Altieri, M.A. (1988) Influences of plant-produced allelochemicals on the host/prey selection behavior of entomophagous insects. In P. Barbosa and D.K. Latournean (eds.), *Novel Aspects of Insect–Plant Interactions*, Wiley, New York, pp. 65–90.

Nordlund, D.A., Chalfont, R.B. and Lewis, W.J. (1994) Arthropod populations, yield and damage in monocultures and polycultures of corn, bears and tomatoes. *Agric. Ecosyst. Environ.*, **11**, 353–367.

Paul, A.V.N. and Yadav, B. (2003) Chemical mediated tritrophic interactions and their implications in biological control. In O. Koul, G.S. Dhaliwal, S.S. Marwaha and J.K. Arora (eds.), *Biopesticides and Pest Management*, vol. 1, Campus Books International, New Delhi, pp. 222–257.

Polis, G.A. (1991) Complex trophic interactions in deserts: an empirical critique of food web theory. *Am. Nat.*, **138**, 123–155.

Pyke, B., Rice, M., Sabine, B. and Zaluki, M. (1987) The push–pull strategy — behavioural control of *Heliothis. Aust. Cotton Grower*, **1987**, 7–9.

Ravensberg, W.J. (1994) Biological control of pests: current trends and future prospects. In *Brighton Crop Protection Conference, Pests and Diseases*, vol. 2, Brighton Crop Protection Council, Farnham, UK, pp. 591–600.

Rosenheim, J.A. (1998) Higher-order predators and the regulation of insect herbivore populations. *Annu. Rev. Entomol.*, **43**, 421–447.

Rosenheim, J.A., Kaya, H.K., Ehler, L.E., Marois, J.J. and Jaffee, B.A. (1995) Intraguild predation among biological control agents: theory and evidence. *Biol. Cont.*, **5**, 303–335.

Singh, S.P. (2001) augmentative biocontrol in India. In S.P. Singh, S.T. Murphy and C.R. Ballal (eds.), *Augmentative Biocontrol*, Proc. ICAR-CABI Workshop, Project Directorate of Biological Control, Bangalore and Punjab Agricultural University, Ludhiana, India, pp. 1–20.

Singh, S.P. (2003) Role of predators and parasitoids in biological control of crop pests. In O. Koul, G.S. Dhaliwal, S.S. Marwaha and J.K. Arora (eds.), *Biopesticides and Pest Management*, vol. 1, Campus Books International, New Delhi, pp. 196–221.

Singh, S.P., Murphy, S.T. and Ballal, C.R. (eds.) (2001) *Augmentative Biocontrol*, Proc. ICAR-CABI Workshop, Project Directorate of Biological Control, Bangalore and Punjab Agricultural University, Ludhiana, India.

Smith, S.M. (1996) Biological control with *Trichogramma*: Advances, successes and potential in IPM. *Annu. Rev. Entomol.*, **47**, 375–406.

Stiling, P. (1993) Why do natural enemies fail in classical biological control programs? *Am. Ent.*, **39**, 31–37.

Sullivan, D.J. (1987) Insect hyperparasitism. *Annu. Rev. Entomol.*, **32**, 49–70.

Sullivan, D.J. and Völkl, W. (1999) Hyperparasitism: multitrophic ecology and behaviour. *Annu. Rev. Entomol.*, **44**, 291–315.

van Emden, H.F. (1989) *Pest Control*, Edward Arnold, London.

Völkl, W. (1997) Interactions between ants and aphid parasitoids: patterns and consequences for resource utilization. *Ecol. Stud.*, **130**, 225–240.

Whitfield, J.B. (1998) Phylogeny and evolution of host–parasitoid interactions in Hymenoptera. *Annu. Rev. Entomol.*, **43**, 129–151.

Whitman, D.W. and Nordlund, D.A. (1994) Plant chemicals and the location of herbivorous arthropods by their natural enemies. In T.N. Ananthakrishnan (ed.), *Functional Dynamics of Phytophagous Insects*, Oxford and IBH Publishing, New Delhi, pp. 133–159.

Williams, T. (1991) Host selection and sex ratio in a heteronomous hyperparsitoid. *Ecol. Entomol.*, **16**, 377–386.

Wright, D.J. and Verkerk, R.H.J. (1995) Integration of chemical and biological control systems for arthropods: evaluation in a multitrophic context. *Pestic. Sci.*, **44**, 207–218.

Yasuda, H. and Kimura, T. (2001) Interspecific interactions in a tritrophic arthropod system: effects of spiders on the survival of larvae of three predatory lady birds in relation to aphids. *Entomol. Exp. Appl.*, **98**, 17–25.

NATURAL ENEMIES AND PEST CONTROL: AN INTEGRATED PEST MANAGEMENT CONCEPT

2

Juan A. Morales-Ramos and M. Guadalupe Rojas
Southern Regional Research Center, United States Department of Agriculture, New Orleans, USA

Introduction

The use of natural enemies has been part of integrated pest management (IPM) from the beginning (Horn, 1988; Luckman and Metcalf, 1994). Because the basic concept of pest management involves the understanding and manipulation of the agroecosystem to reduce pest outbreaks, the natural enemy complex takes a central role in the application of IPM. Kogan (1998) defines a continuum of IPM adoption divided into three levels, where the minimum set of tactical components to qualify as IPM includes field scouting for pests and natural enemies, inaction thresholds, crop rotation, and use of selective pesticides. From these tactics, the use of inaction thresholds and natural enemy counts is compatible with natural enemy conservation. Biological control by augmentation is considered an important tactic within the first level of IPM integration (Kogan, 1998).

The equilibrium population size and dynamic behavior of many phytophagous insects are largely determined by their natural enemies, which are key factors keeping many phytophagous insects rare (Waller, 1987). Reduction of natural enemy pressure is one of the most important factors in the development of insect outbreaks (Andrewartha and Birch, 1982). Density-dependent factors such as natural enemies tend to regulate herbivorous populations under conditions of abundance (Huffaker *et al.*, 1971). Disruption of the natural enemy complex by application of insecticides or by other adverse factors usually results in insect outbreaks (Risch, 1987).

The most notorious cases of insect outbreaks induced by natural enemy reduction are those in which an imported pest escapes its natural enemies by colonizing a new environment (Risch, 1987). Horn (1988) estimates that 40 percent of the most damaging agricultural pests in North America originated elsewhere. Indigenous insects become pests when they colonize an introduced crop (Risch, 1987). Often, herbivorous insects can colonize a new host plant and become adapted to it more easily than can their natural enemies. The boll weevil, *Anthonomus grandis* Boheman, is a good example, switching from its original host plant complex (*Hampea* spp.) to cotton upon its cultivation in Meso-America (Burke *et al.*, 1986). The boll weevil escaped its natural enemies in two ways: (i) because of the switch of host plant, most of the boll weevil parasitoids were unable to recognize cotton as the host habitat; and (ii) through migration north, the tropical natural enemy complex was unable to adapt to hard winters because of the inability to overwinter (King *et al.*, 1996).

The concept of IPM

In its early origins, integrated control or integrated pest management was conceived with biological control as the central tactic to minimize the use of insecticides (Stern *et al.*, 1959). Insect pest management and integrated control were the origins of modern IPM (Frisbie and Adkisson, 1985; Horn, 1988). Pest management is defined as the use of all available techniques in a unified program to manage pest populations, avoiding economic damage and minimizing environmental effects (NAS, 1969). Under the concept of pest management, the pest status is viewed in a more tolerant manner and coexistence with insect pests causing some damage becomes acceptable (Luckman and Metcalf, 1994). Integrated pest management (IPM) is defined as a pest population management system that utilizes all suitable techniques in a compatible manner to reduce pest populations and maintain them at levels below those causing economic injury (Frisbie and Adkisson, 1985). IPM is also defined as the intelligent selection and use of pest control tactics that will ensure favorable economic, ecological, and sociological consequences (Luckman and Metcalf, 1994). Kogan (1998) defines IPM as a decision support system for the selection and use of pest control tactics, singly or harmoniously coordinated into a management strategy, based on cost–benefit analyses that take into account the interest of and impacts on producers, society, and the environment.

Hoy (1994) emphasizes that biological control should play a central role in IPM, thus eliminating reliance on pesticides. Unfortunately, biological control is more often a minor component of IPM systems (Hoy, 1994). There are three main reasons for the limited use of biological control in IPM systems. The first is the belief that biological control is incompatible with the use of pesticides. The second concerns the perception that pests of annual crops are not amenable to biological control. However, the most important reason is that pesticides are believed by growers to offer a short-term, uncomplicated solution to pest problems, while biological control requires knowledge and understanding of the agroecosystem by the user if it is to be

Juan A. Morales-Ramos and M. Guadalupe Rojas

applied effectively. This chapter will present evidence to contradict these beliefs and present a new context for the application of biological control.

Biological control approaches

DeBach (1964) defined biological control as the use of parasites, predators, and pathogens to maintain insect population densities below economic damage levels. A later definition by Van Driesche and Bellows (1996) adds the use of antagonist and competitor species to the definition. In these authors' view, biological control is a process in which one species population lowers the numbers of another species by mechanisms such as predation, parasitism, pathogenesis, or competition. Cate (1990), in an attempt to unify the concept of biological control in all disciplines, defines biological control as a unique pest control strategy that seeks to reestablish homeostasis in the interactions among organisms disrupted by agricultural activities.

Success in biological control is measured as complete, substantial, or partial according to the extent of population regulation achieved (DeBach, 1964). Complete success is achieved when the populations of the target pest are maintained below economic damage levels in an extensive area so that pesticide treatments become rarely, if ever, necessary. Substantial success is achieved if only occasional pesticide treatments are required to maintain the pest population below economically damaging levels. Partial success refers to cases in which pesticide application remains necessary but less frequently or where complete success is achieved only in a minor portion of the pest-infested area (DeBach, 1964; DeBach et al., 1971).

There are three basic tactics used in biological control: importation, conservation, and augmentation (DeBach, 1964; Hoy 1994; Van Driesche and Bellows, 1996). These tactics, although they may be successful individually, are only tools in a continuum of biological control efforts and they frequently interact within a pest control program (Figure 2.1). For example, conservation efforts often follow introduction or augmentation efforts, and augmentation may be part of the conservation efforts or may be applied after a partially successful introduction (Figure 2.1). In this chapter we will emphasize the concept of biological control as a single approach that integrates these three tactics.

Importation, or classical biological control, consists of the foreign exploration for exotic natural enemies, their importation, and their release to control a pest that has been accidentally or erroneously introduced (DeBach, 1964; Caltagirone, 1981; Hoy, 1994; Van Driesche and Bellows, 1996). Numerous successes of biological control by importation have been documented (DeBach, 1964; DeBach et al., 1971; Caltagirone, 1981; Luck, 1981; Hoy, 1994); however, this approach has been applied to a very small proportion of the world's pest species (Van Driesche and Bellows, 1996). The success of natural enemy importation has been estimated to result in approximately 17 percent complete and another 43 percent substantial or partial successes, using data from 1200 projects (Van Driesche and Bellows, 1996). However, the successes of importation have been substantially higher when directed against pests of perennial crops (22 percent complete and 72 percent substantial or partial) as compared to those obtained in annual crops (3 percent complete and 43 percent substantial or

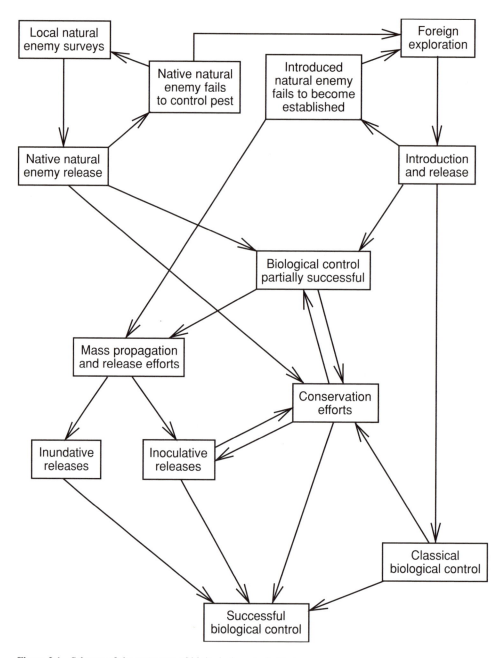

Figure 2.1 Scheme of the sequence of biological control efforts.

Juan A. Morales-Ramos and M. Guadalupe Rojas

partial) (Greathead, 1986). Importation alone does not constitute a highly effective approach against all pests in different environments but, as part of the continuum (see Figure 2.1), it constitutes the first and one of the most important stages of successful biological control.

Conservation is the most frequently used biological control tactic in IPM and is defined as actions to preserve and increase natural enemies by environmental manipulation (DeBach, 1964; Ehler, 1998). To be effective, the conservation tactic requires the presence of appropriate natural enemies within the agroecosystem. Natural enemies must first be introduced in cases in which this condition is not met (Van Driesche and Bellows, 1996). Ehler (1998) does not differentiate between conservation and augmentation and considers both tactics to be along a continuum from selective use of pesticides to inundative releases.

Augmentation, usually considered as the tactic of last resort in biological control, has been successfully applied in cases in which all other tactics have failed. Augmentation is defined as the effort to increase populations of natural enemies either by propagation and release or by environmental manipulation (DeBach, 1964; Rabb et al., 1976). King (1993) restricts this definition to the efforts of mass propagation and release of natural enemies. The tactics of biological control by augmentation have two approaches: inoculative and inundative releases (DeBach, 1964). These approaches have different applications depending on the adaptability of the natural enemy and the environment in which it must operate. Inoculative releases are used in cases in which the control is expected to result from the progeny of the released natural enemies. Inundative releases are used in cases in which control is expected through the direct activities of the natural enemies released (DeBach, 1964; Parrella et al., 1992).

The three tactics of biological control are not always effective individually; however, when they are viewed as stages of biological control efforts and used in an integrated manner, they can be a powerful tool in IPM (Figure 2.1). Introduction of natural enemies should be the first step when no adequate natural enemies are present. Conservation can substantially improve the effectiveness of importation or extend its range of success. Augmentation can be used to attack pests of annual crops or in environments not favorable for establishment of natural enemies or rapid numerical response. The impact of augmentation can be improved by conservation. Often, augmentation can aid the success of importation; for instance, the percentage of successful control by introduction is positively correlated with the number of individuals released and the number of releases (Greathead, 1986).

Use of natural enemies under the context of IPM

By definition, IPM requires the application of the most effective, economic, and environmentally friendly tactics to control a pest. Therefore, biological control should be among the first tactics to be explored in an IPM program.

Biological control by importation of natural enemies

Despite the success of classical biological control, seldom is complete control of a pest achieved by this method (Greathead, 1986; Hoy, 1994). In fact, the concept of integrated control was born from the need to include other control methods after the importation of natural enemies resulted in only partial control of the spotted alfalfa aphid, *Therioaphis trifolii* (Monell), during the 1950s (Caltagirone, 1981).

The successful introduction of the parasitoid *Pediobius foveolatus* (J.C. Crawford) from India against the Mexican bean beetle, *Epilachna varivestis* Mulsant, in the US Atlantic coastal plains is an example of a successful classical biological control within an IPM program in soybeans (Kogan and Turnipseed, 1987).

Biological control by importation has been an important part of the IPM efforts against the alfalfa weevil, *Hypera postica* (Gyllenhal) (Flanders and Radcliffe, 1999) with the successful establishment of *Bathyplectes curculionis* (Thompson) and *B. anurus* (Thompson) (Kingsley *et al.*, 1993; Steffey *et al.*, 1994). A national program in the USA (USDA, ARS) to introduce natural enemies of the alfalfa weevil achieved the establishment of at least five hymenopteran parasitoids (Radcliffe and Flanders, 1998). Substantial control of the alfalfa weevil has been attributed to the activities of five key natural enemies: the ichneumonid larval endoparasitoids *B. curculionis* and *B. anurus*, the braconid adult endoparasitoid *Microctonus aethiopoides* Lioan; the eulophid larval endoparasitoid *Oomyzus incertus*; and the fungal pathogen *Zoophthora phytonomi* (Radcliffe and Flanders, 1998). The introduction of *Z. phytonomi* is believed to have been accidental and questions concerning its origin have not been resolved (Radcliffe and Flanders, 1998). Since the introduction of the program, Kingsley *et al.* (1993) have reported a significant decrease in the use of insecticides in alfalfa. However, these reductions could not be directly attributed only to the biological control success against the alfalfa weevil, but also to a general awareness of the benefits of conservation (Kingsley *et al.*, 1993).

Even after a complete success has been achieved through importation of natural enemies, conservation efforts must continue to ensure effective biological control. In 1947 the use of DDT for citrus pest control in California was followed by outbreaks of the cottony cushion scale, *Icerya purchasi* Maskell, which had been under complete biological control since the introduction of *Rodolia cardinalis* (Mulsant) (Metcalf, 1994). In 1981 the use of malathion to control the Mediterranean fruit fly, *Ceratitis capitata* (Wiedemann), in California triggered outbreaks of the olive scale, *Parlatoria oleae* (Colvee), by eliminating introduced parasitoids, which had kept the scale under complete biological control (Ehler and Endicott, 1984).

Biological control by conservation of natural enemies

The importance of natural enemies in the regulation of pest populations has been demonstrated (Huffaker *et al.*, 1971; Price, 1987; Van Driesche and Bellows, 1996). Pest outbreaks following the application of broad-spectrum insecticides have been considered as evidence of the importance of natural enemies in pest population regulation (Price, 1987). Efforts to control the boll weevil, *A. grandis*, in northeastern

Mexico's Rio Grande Valley by the use of broad-spectrum insecticides during the 1960s resulted in catastrophic outbreaks of *Helicoverpa zea* (Boddie) and *Heliothis virescens* (Fabricius) that ended in the decline of the cotton industry in the region (Adkisson, 1971; Metcalf, 1994). The sweetpotato whitefly, *Bemisia tabaci* (Gennadius), became a destructive pest after the introduction of pyrethroids in California's Imperial Valley in 1975 (Metcalf, 1994). Wilson *et al.* (1998) demonstrated that outbreaks of spider mites in cotton in Australia were correlated with the elimination of natural enemies by early-season insecticide usage.

During the late 1980s, the use of malathion to control the boll weevil in the southeastern United States resulted in uncharacteristic outbreaks of the beet armyworm, *Spodoptera exigua* (Hubner), a secondary pest of cotton (Ruberson *et al.*, 1998). Studies have demonstrated that a complex of natural enemies is capable of maintaining beet armyworm populations below economic damage levels and that this complex is easily disrupted by the use of broad-spectrum insecticides, especially organophosphates, triggering beet armyworm outbreaks (Ruberson *et al.*, 1994). Similarly, beet armyworm outbreaks reported in the Rio Grande Valley of Texas during the 1995 growing season were correlated with area-wide applications of ULV malathion (Summy *et al.*, 1996). This outbreak was also correlated with a decrease in densities of natural enemies in Texas fields compared to cotton fields in the Mexican side of the Rio Grande Valley, where no outbreaks occurred and no area-wide applications of malathion were made (Summy *et al.*, 1996). During the following season, area-wide applications of malathion were suspended in the Rio Grande Valley and studies of beet armyworm populations during that year showed that natural enemies, including predaceous Heteroptera, Coccinellidae, and green lacewings, maintained this pest under control even within insecticidal check fields (Summy *et al.*, 1997).

Conservation of natural enemies is the most common link between biological control and other control practices in an IPM system. The most important ways to conserve natural enemies within an agroecosystem are cultural control by habitat manipulation and rational use of pesticides (Hoy, 1994; Van Driesche and Bellows, 1996).

Habitat manipulation

Modern crop systems are often inhospitable for the proliferation of natural enemies. Habitat manipulation can enhance natural enemy survival by providing alternative hosts or prey, providing food in the form of pollen and/or nectar for parasitoids, refugia from winter or higher-order predators, and a continuous source of primary host or prey (Powell, 1986). Habitat instability has been one of the main reasons for the failure of imported natural enemies to become established in a new environment (Hoy, 1985). Special efforts in cultural practices and habitat manipulation, particularly in monoculture annual crops, may be required to achieve establishment of natural enemies (Hoy, 1985). Kogan and Lattin (1993) suggested that generalist predators are the functional group most likely to benefit from conservation efforts within an IPM program and that such efforts should emphasize the increase in plant and insect diversity within the agroecosystem.

It is imperative to know the biology of the natural enemies to be able to make advantageous modifications of their environment. However, the complexity of agricultural ecosystems demands fundamental studies of crop systems under relevant field conditions for the selection of the most appropriate manipulation strategies (Verkerk *et al.*, 1998).

Parasitoids and predators often have biological requirements that are not provided by their host or prey environment (Barbosa and Bentry, 1998). Increasing diversity in time and/or space within the agroecosystem by multiple cropping, intercropping, strip harvesting, selective retention of weeds, or conservation of wild plants at field margins helps the conservation of natural enemies by satisfying their biological requirements (Powell, 1986). Adult females and males of many parasitoids feed on pollen, sugary plant juices, and nectar of flowers (House, 1977). The importance of carbohydrate sources for the longevity and fecundity of many parasitoids has long been recognized (Flanders, 1935; Edwards, 1954; Leius, 1961; Bracken, 1965; Lum, 1977). The lack of food sources for adult parasitoids in the field was a possible reason for the failure of many early attempts at biological control (Beirne, 1962). Establishment of hymenopterous parasitoids depends on the presence of suitable species of flowers that provide nutrients to adult parasitoids (Leius, 1960). Syme (1975) found that the presence of wild flowers significantly increased the longevity of two parasitoids of *Rhyacionia buoliana* (Denis & Schiffermuller), on the pine shoot. Some parasitoids are attracted to flowers and other plant parts, which may provide food. *Campoletis sonorensis* (Cemeron), a parasitoid of *H. virescens*, showed attraction to flowers of five different plants including tobacco, cotton, sorghum, bluebonnet, and geranium (Elzen *et al.*, 1983). Wild *Brassica vulgaris* and *B. kaber* in the vicinity of cabbage fields may enhance the survival of the ichneumonid *Diadegma insularis* (Cresson), a parasitoid of the diamondback moth, *Plutella xylostella* (Linnaeus), by providing floral nectar reserves (Idris and Grafius, 1996). Border plantings of *Phacelia tanacetifolia* increased the abundance of syrphid flies in cabbage fields by providing pollen and resulted in reduced aphid damage in cabbage (White *et al.*, 1995).

Some parasitoids do not use floral nectar to fulfill their nutritional requirements. The eulophid, *Edovum puttleri* Grissell, a parasitoid of Colorado potato beetle eggs, feeds on the honeydew produced by aphids and other homopterans (Idoine and Ferro, 1988). In situations in which the available flower species are unsuitable, additional carbohydrate sources may be provided to enhance the survival of natural enemies. The use of food sprays composed of hydrolyzed protein, brewers' yeast, and carbohydrates has been successful in increasing the densities of predaceous insects in alfalfa and cotton fields (Hagen *et al.*, 1971).

The choice of a particular crop variety may be an important factor for conservation of natural enemies. The fecundity, survival, and parasitism rate of *C. sonorensis* were lower in nectariless than in nectaried cotton varieties owing to the lack of extrafloral nectaries that provide needed carbohydrates (Lingren and Lukefahr, 1977). The use of nectariless cotton varieties resulted in a significant reduction of natural enemy densities (Adjei-Maafo and Wilson, 1983). Pea variety significantly affected the parasitism rate on pea weevils, *Bruchus pisorum* (Linnaeus), by the pteromalid parasitoid *Eupteromalus leguminis*, and the parasitism rates were negatively correlated with seed infestation by weevils (Annis and O'Keeffe, 1987).

Diversification within the crop system by intercrossing may increase the density and diversity of natural enemies. Corbett and Plant (1993), using a simulation model, concluded that interplanting might increase the number of highly mobile natural enemies. Sesame interplantings in cotton provided a suitable environment to increase the diversity and density of *Heliothis* spp. parasitoids in Mississippi (Pair *et al.*, 1982). The larval parasitism of the European corn borer, *Ostrinia nubilalis* (Hubner), was significantly higher in cornfields near woodlands than in fields next to open spaces or in field interiors (Landis and Haas, 1992). The presence of prune trees within vineyards significantly increased densities of *Anagrus epos* Girault, a mymarid egg parasitoid of the western grape leafhopper *Erythroneura elegantula* Osborn, in California (Murphy *et al.*, 1996). Prune trees provided overwintering sites for *A. epos* and also provided a friendlier environment for the parasitoids by their wind-breaking effect (Murphy *et al.*, 1996).

The use of resistant plant varieties tends to increase the effectiveness of parasitoids against phytophagous insects by increasing their developmental time and thereby increasing vulnerability to larval parasitoids. The parasitism of fall armyworm, *Spodoptera frugiperda* (J.E. Smith), was significantly higher in resistant than in susceptible varieties of corn (Riggin *et al.*, 1992). Incorporation of *Bacillus thuringiensis* toxins in genetically engineered tobacco increased parasitism of *H. virescens* by *C. sonorensis* (Johnson and Gould, 1992).

Conservation-tillage practices increase the densities of some predatory insects. Significantly higher populations of spiders and carabid beetles have been reported in reduced-tillage and no-tillage than in conventionally plowed crops (Stinner and House, 1990). Greater damage by cutworms was reported in corn in a conventional than in a no-tillage system in Ohio, and this damage was negatively correlated with the absolute density of predators (Brust *et al.*, 1985, 1986). The effectiveness of the parasitoid *Catolaccus grandis* (Burks) against the boll weevil, *A. grandis*, in cotton is greatly reduced by mechanical cultivation, which buries fallen infested cotton buds and protects the immature weevils from parasitism (Summy *et al.*, 1994a).

The use of mulch helps increase the survival of predators by providing moisture and shelter. The use of straw-mulch in potato fields significantly increased the densities of soil-inhabiting predators and reduced defoliation damage by Colorado potato beetles (Brust, 1994). The mulch initially induced an increase in carabid predators early in the season, but by June and July the predators *Coleomegilla maculata* (DeGeer), *Hippodamia convergens* (Guerin-Meneville), *Chrysoperla carnea* (Stephens), and *Perillus bioculatus* (Fabricius) comprised 75 percent of the predators found in foliage and accounted for 86 percent of the observed predation (Brust, 1994).

Conservative use of pesticides

Within an IPM system, insecticides should be used only when pest populations exceed a specific level and when no other control tactic is available (Hoy, 1994). Conservation of natural enemies can be compatible with the use of pesticides by use of chemicals that are more selective, less toxic to natural enemies, with lower residual activity, applied in minimal dosages, and timed and placed with the pest population (Metcalf, 1994). Selectivity of pesticides can be classified as physiological or ecological, where physiological selectivity is defined as differential toxicity between animal taxa due to the organism's biochemical process (Hull and Beers, 1985; Metcalf, 1994). Ecological selectivity is the judicious use of pesticides, based on critical selection, timing, dosage, placement, and formulation of pesticides with the goal of maximizing pest mortality while minimizing beneficial arthropod mortality (Hull and Beers, 1985).

Some pesticides have a selective toxicity to the target pest, leaving their natural enemies mostly unaffected. This differential toxicity is usually common in pesticides that require ingestion by the target pest. The pesticide abamectin, a natural product derived from the actinomycete *Streptomyces avermitilis,* was found to be selective against the greenhouse whitefly, *Trialeurodes vaporariarum* (Westwood), causing minimal mortality to its parasitoid *Encarsia formosa* Gahan (Zchori-Fein *et al.,* 1994). Studies revealed that toxicity of abamectin to adult *E. formosa* disappeared 24 hours after application as the plants absorbed the chemical. Zchoiri-Fein *et al.* (1994) theorize that *T. vaporariarum* susceptibility is due to its feeding on treated plants and parasitoid survival on whitefly pupae is explained by the lack of plant-feeding during this developmental stage. The effects of this type of chemical may vary between different natural enemy species. The growth regulator teflubenzuron, used to control the diamondback moth, *P. xylostella,* in Malaysia had no effect on parasitism rates of this pest by *Cotesia plutellae* but significantly reduced parasitism by *Diadegma semiclausum* Hellen (Furlong *et al.,* 1994).

Determining the minimal dose of a pesticide needed to control a pest and reducing applications to this minimum often contributes substantially to natural enemy conservation. The use of methyl parathion and methomyl in combination produced outbreaks of lepidopterous defoliators in soybean and was correlated with the reduction in densities of natural enemies (Shepard *et al.,* 1977). Similarly, Morrison *et al.* (1979) observed an 8-fold increase in populations of *H. zea* in soybean crops after application of methyl parathion or aldicarb. Current pesticide use in soybean is based upon selective strategies consisting of application of minimal dosages least detrimental to natural enemies (Kogan and Turnipseed, 1987).

Timing of pesticide treatments is often the most effective and economical method of achieving selectivity of pesticides while conserving natural enemies (Hull and Beers, 1985). Wilson *et al.* (1998) determined that natural enemies play an important role in the population dynamics of *Tetranychus urticae* Koch, a key pest of cotton in Australia. The selection of pesticides used early in the season to control other pests had a significant effect on the survival of predator populations and on the outbreaks of *T. urticae* (Wilson *et al.,* 1998). Murchie *et al.* (1997) concluded that the detrimental effect of insecticides on the parasitism of *Ceutorhynchus assimilis* (Paykull) by the parasitoid *Trichomalus perfectus* could be minimized by timing the applications before parasitoid migration to the fields, which occurs 3–4 weeks after *C. assimilis* migration. Rapidly degradable pesticides, applied in a timely manner, are more suitable for minimizing pesticide effects on natural enemies. The use of a single spray of methyl parathion or malathion in late March to control overwintering alfalfa weevils, *H. postica,* allows the parasitoid *B. curculionis* to thrive later because those insecticides degrade before the emergence of this parasitoid (Wilson and Armbrust, 1970). The use of early-season sprays of methyl parathion was successfully integrated with augmentative releases of the pteromalid parasitoid, *Catolaccus grandis* (Burks) to control the boll weevil, *A. grandis,* in West Alabama (Morales-Ramos *et al.,* 1994). Rapid degradation of this pesticide allows safe timing of sprays to control overwintering boll weevils and fleahoppers before releases of *C. grandis* are required to control F_1 and F_2 boll weevil larvae and pupae (King *et al.*, 1993; Summy *et al.*, 1994b).

The method of pesticide application is an important factor in natural enemy conservation. For instance, aerial applications often inefficiently deliver only a fraction of the pesticide to the target pest, yet spread pesticides kilometers away by wind-induced spray drifts (Currier *et al.*, 1982; Hull and Beers, 1985; Metcalf, 1994). Ground sprayers provide better accuracy and control of delivery of pesticides, allowing their strategic placement within the field. Restricting the pesticide to a specific plant part or field location minimizes its impact on beneficial organisms (Hull and Beers, 1985).

Another important tool of conservation concerning selective use of pesticides is the nonintervention decision tactic. This tactic consists of monitoring natural enemy densities and understanding their impact on pest populations to make decisions on the use of pesticides (Sterling *et al.*, 1989). The use of simulation models to assess the impact of key pest mortality factors operating in the field can aid the decision-making process and effectively reduce the number of insecticide applications. For instance, the mortality factors affecting boll weevil populations in Texas have been studied extensively (Curry *et al.*, 1982; Fillman and Sterling, 1983; Sturm and Sterling, 1986; Sturm *et al.*, 1989, 1990). It has been documented that densities of 0.4 red imported fire ants, *Solenopsis invicta* Buren, per plant provide boll weevil control about 90 percent of the time (Fillman and Sterling, 1985). Eliminating the use of pesticides helps to conserve populations of natural enemies that maintain control of secondary pests 95 percent of the time in Texas and California (Sterling *et al.*, 1989).

Biological control by augmentation of natural enemies

Augmentation of natural enemies is the biological control tactic that requires the greatest amount of effort and resources for its implementation. Augmentation of natural enemies requires the use of mass-rearing techniques (mass propagation) to produce natural enemies in large numbers and mass-handling techniques for distribution and release of these natural enemies in the field (DeBach and Hagen, 1964; Stinner, 1977; King *et al.*, 1985). Such factors make the use of augmentation tactics costly and many times economically unfeasible (King *et al.*, 1985; King and Powell, 1992; Parrella *et al.*, 1992). For this reason, augmentation efforts should not be considered until all other biological control tactics have been attempted and have met with failure or only partial success.

Augmentation of natural enemies provides a biological solution to pest problems in ephemeral crops where naturally occurring beneficial organisms fail to respond quickly enough to control populations of primary pests (King, 1993). The natural enemy complex in an agroecosystem has a density-dependent effect on pest populations (Huffaker *et al.*, 1971). This density dependence means that more often than not the natural enemy complex will be unable to respond numerically to a rapidly increasing pest population within an annual crop system. Knipling (1977) used a simple mathematical model to predict that augmented natural enemies would operate in a density-independent manner, inducing the crash of the target pest population. A simulation model developed independently to study the effects of inundative releases of *C. grandis* on boll weevil populations predicted similar density independence of such systems (Morales-Ramos *et al.*, 1996). The predictions of this model were confirmed by experimental inundative releases of *C. grandis* in cotton fields in the Rio Grande Valley (Summy *et al.*, 1995; Morales-Ramos *et al.*, 1995).

The impact of predator and parasitoid interactions on a pest population is measured in terms of their functional and numerical responses (Huffaker *et al.*, 1971). While functional response depends mainly on the searching capacity of the parasitoid or predator, the numerical response of natural enemies tends to be dependent on host or prey populations and is always delayed at least one generation, depending on the developmental time of the beneficial insect (Huffaker *et al.*, 1971). Murdock and Oaten (1975) stated that predator–prey system stability and population regulation of a pest depend on multiple species interactions and spatial heterogeneity. This situation occurs within perennial crops where diversity is maintained. Annual crops, however, tend to be highly homogeneous and exhibit low diversity, which is rarely maintained after the end of the crop season. As a result, annual crops behave like islands that are colonized by arthropods from surrounding areas (Price and Waldbauer, 1994). These crops tend to be colonized first by *r*-strategists, highly mobile species with high reproductive rates (Price, 1984). Their predators and parasitoids are slower in colonizing new environments (Price and Waldbauer, 1994).

Augmentation of natural enemies provides a solution to annual crop vulnerability to outbreaks of colonizing *r*-strategists by providing an artificially induced numerical response of a key natural enemy. The instability of these systems does not present a problem for augmentation tactics because establishment of the augmented natural enemy is not required (King *et al.*, 1996). Augmentation efforts are most beneficial when directed at the control of primary pests, with *r*-strategist characteristics, of ephemeral crop systems (Price, 1992; King, 1993).

The cost of mass propagation and handling of natural enemies is the most important limitation for the use of the augmentation approach, and the feasibility of mass propagation is a prerequisite and often the main bottleneck for implementing augmentative biological control (King *et al.*, 1985; King and Powell, 1992; King, 1993). One of the recommendations for the successful implementation of natural enemy augmentation was the development of artificial diets for rearing parasitoids and predators (King *et al.*, 1985).

The development of artificial diets for insect parasitoids has been advancing consistently. During the mid-1980s, most parasitoids reared on artificial diets developed to adulthood but with serious reductions in fitness (Thompson, 1986). A recent review shows more significant advances, with parasitoids reared *in vitro* having fitness similar to or improved over that of those reared on natural hosts (Thompson, 1999). Thompson and Hagen (1999) list 35 species of dipterous and hymenopterous parasitoids that have been reared successfully to the adult stage on artificial diets, but only 18 species have been reared on meridic diets (lacking insect components). Some examples include a meridic diet for the boll weevil parasitoid, *C. grandis* (Rojas *et al.*, 1996), which produced females with fecundity and searching ability close to those reared *in vivo* (Morales-Ramos *et al.*, 1998). A meridic diet containing tissue culture media for the *Spodoptera* spp. parasitoid *Diapetimorpha introita* (Cresson) produced parasitoids with biological parameters approaching those of host-reared parasitoids (Carpenter and Greany, 1998). However, all the successes in *in vitro* parasitoid rearing have been achieved with idiobiontic parasitoids and no successes have been reported for koinobiontic Hymenoptera (Thompson, 1999). The hosts of idiobiontic parasitoids do not continue development after parasitism, essentially stopping all physiological processes, while the hosts of koinobiontic parasitoids continue their development, making hormonal interactions an important part of the host–parasitoid relationship (Quicke, 1997). Rearing koinobiontic parasitoids *in vitro* is not a purely nutritional problem and requires the understanding of complex physiological and nutritional interactions between host and parasitoid.

A better success rate has been obtained in rearing predators on meridic artificial diets (Thompson, 1999). Thompson and Hagen (1999) report 40 species of predator insects that have been reared successfully from egg to adult on artificial diets, but only 9 of these successes were achieved on meridic diets. A few of the most successful examples include the rearing of *C. carnea* (Hasagawa *et al.*, 1989) and *C. rufilabris* (Butmeister) (Cohen and Smith, 1998). The successful rearing of *Geocoris punctipes* (Say) on meridic diets for many generations is one of the most important successes to date (Cohen, 1981, 1985; Cohen and Staten, 1994). The predatory stinkbugs *Podisus maculoventris* (Say) and *P. sagitta* were successfully reared on a similar meat-based diet (De Clercq and Degheele, 1992). More recently *Perillus bioculatus* (Fabricius), a stinkbug predator of the Colorado potato beetle, was successfully reared for 11 generations on a high-protein diet (Rojas *et al.*, 2000).

The effort required for the implementation of an augmentation program demands careful scrutiny of the candidate natural enemy species. Some attributes of natural enemies that have been described by DeBach (1964) and by Messenger and van den Bosch (1971) as advantageous in a classical biological control program are also import-ant in an augmentation program. However, the most important attributes are the ability to locate the host or prey in extremely low population densities and to survive in the pest environment. Unlike importation of natural enemies, augmentation does not require the natural enemies to be able to become established (King, 1993). Some additional advantageous characteristics of natural enemies for augmentation include a life cycle shorter or equal to that of the pest target, high reproductive capacity, host or prey selectivity, and feasibility of mass rearing.

As the technology for mass propagation of natural enemies' progresses, a new industry of biological control develops. A meridic diet for parasitic and predaceous insects (Greany and Carpenter, 1998) and an improved meridic diet for rearing the parasitoid *C. grandis* (Rojas *et al.*, 1999) have been patented. Methods for the mech-anized mass propagation of *C. grandis* are currently under development (Edwards *et al.*, 1998). By 1997, there were approximately 140 species of natural enemies commercially available from independent suppliers in the USA, Mexico, and Canada (Hunter, 1997).

Inoculative releases

If the control of a pest by augmentative releases is expected to come from the progeny or successive generations of the released natural enemies, the releases are defined as inoculative (DeBach, 1964). These efforts are usually directed at pests that have been under partial biological control or at pests under complete biological control that have reappeared because of disruption of the natural enemy. An example is the use of *Aphytis melinus* DeBach, *A. chrysomphali* (Mercet), and *A. lingnanensis* Compere against the California red scale, *Aonidiella aurantii* (Maskell) (DeBach and White, 1960; King and Morrison, 1984). Inoculative releases of the eulophid parasitoid, *Pediobius foveolatus* (J.C. Crawford) have been effective in suppressing the Mexican bean beetle, *Epilachna varivestis* Mulsant, in bean crops in the eastern USA (Flanders, 1985). Inoculative releases of this parasitoid have ensured the effectiveness of this biological control program within an annual crop environment (Kogan *et al.*, 1999).

Inoculative releases of natural enemies are often directed at secondary pests that have become a problem because of disruption by climate or pesticide application. The use of generalized predators is common in such cases. Examples of this include releases of *C. carnea* to control *Heliothis* spp. and other lepidopterous pests in cotton (King *et al.*, 1985), and the release of coccinellid predators to control aphid pests. Natural enemies released in this manner do not require production in very large numbers to be effective because the pest population is controlled by in-field reproduction of the released natural enemies. Therefore, these strategies do not require the development and construction of specialized facilities and equipment.

Despite the success of the parasitoid introductions against the alfalfa weevil, insecticide use has continued to be an important part of insect management in alfalfa (Radcliffe and Flanders, 1998). Augmentation of one or more of the key natural enemies of the alfalfa weevil may further reduce insecticide use in this crop. The parasitoid *Oomyzus* (=*Tetrastichus*) *incertus* (=*erodesi*) is one of the key natural enemies of alfalfa weevil and is a promising candidate for augmentation because it is multivoltine and gregarious (Streams and Fuester, 1967), which are advantageous characteristics for mass propagation. Because *O. incertus* is adapted to high altitudes, its emergence is not synchronized with the highest abundance of alfalfa weevil larvae in spring, resulting in reduced impact (Radcliffe and Flanders, 1998). The importance of *O. incertus* in regulating alfalfa weevil populations depends in large part upon the extent to which fall and summer weevil larvae contribute to adult weevil densities the following spring (Streams and Fuester, 1967). Inoculative releases could be effectively timed to inflict the highest impact on alfalfa weevil populations, thus increasing the regulating effect of *O. incertus*.

Inundative releases

Augmentative releases are defined as inundative when the control of the target pest is expected to be achieved by the direct activities of the released natural enemies (DeBach, 1964). The use of inundative releases of natural enemies is often directed to primary pests of annual or perennial crops that have not been successfully controlled by established natural enemies or introductions.

Some of the most successful uses of inundative releases of natural enemies have been applied to greenhouse crops, specifically releases of *Encarsia formosa* Gahan against the greenhouse whitefly, *Trialeurodes vaporariorum* (Westwood) (King *et al.*, 1985; van Lenteren *et al.*, 1996). The uses of *E. formosa* and *Eretmocerus eremicus* have also been successful against the silverleaf whitefly, *Bemisia argentifolii* Bellows & Perring, in greenhouse poinsettia (Hoddle *et al.*, 1997, 1998, 1999). The most widespread uses of inundative releases have been with *Trichogramma* spp. against several different pests around the world. The use of *Trichogramma* in augmentative releases has been reviewed extensively (King *et al.*, 1985; King, 1993; Smith, 1996; Elzen and King, 1999).

Some promising applications of augmentation by inundative releases include the use of the pteromalid parasitoids *Muscidifurax raptor* Girault & Sanders, *M. zaraptor* Kogan & Legner, *Spalangia endius* Walker, *S. nigroaenea* Curtis, and *S. cameroni* Perkins to control muscoid flies in poultry houses and cattle feedlot environments (Axtell and Rutz, 1986; Meyer, 1986; Patterson and Morgan, 1986). Some of these parasitoids have been successful in some environments and have failed in others (Axtell and Rutz, 1986). Selection of the most suitable species to release in a given environment is one of the main considerations for the application of inundative releases to control muscoid flies (Meyer, 1986). Considerable advances have been achieved in the development of mass propagation systems for these pteromalid parasitoids (Morgan, 1986). *S. cameroni, S. endius, S. nigroaenea, M. raptor, M. zaraptor*, and *M. raptorellus* are commercially available in North America (Hunter, 1997). In addition, the use of artificial diets to rear *Muscidifurax* spp. is promising, these having been successfully reared from egg to fecund adult (M.G. Rojas unpublished) in a modification of the *C. grandis* diet developed by Rojas *et al.* (1996).

The success of inundative releases of *C. grandis* against the boll weevil in cotton has been demonstrated in experimental and commercial cotton fields (Summy *et al.*, 1995; Morales-Ramos *et al.*, 1994, 1995; King *et al.*, 1995). As a key pest of an annual crop (Frisbie *et al.*, 1994), the boll weevil has not been considered amenable to biological control. Efforts to control the boll weevil by the classical approach were not successful (Johnson *et al.*, 1973) and natural control is not effective in most of the cotton-producing areas (Chesnut and Cross, 1971; Sturm *et al.*, 1989). The parasitoid *C. grandis* was selected from a list of introduced parasitoids from importation efforts by Cate *et al.* (1990). This parasitoid was selected on the basis of its attributes, which included high searching capacity (Morales-Ramos and King, 1991), higher rate of population increase than the host, and adaptability to the Texas environment (Morales-Ramos and Cate, 1992). Another important consideration was that a mass-rearing system for the boll weevil was available (Roberson and Wright, 1984) and an encapsulation method in Parafilm® allowed the rearing of *C. grandis* without the need of cotton buds (Cate, 1987), making its mass propagation feasible (Morales-Ramos *et al.*, 1992). Aided by a simulation model, Morales-Ramos *et al.* (1996) established release rates and timing of inundative releases for suppression of the damaging F_1 and F_2 boll weevil generations.

Inundative releases of *C. grandis* have been used in combination with pesticides by timing the applications one week before the first release (Summy *et al.*, 1994b). In a study in commercial cotton in the Rio Grande Valley, inundative releases of *C. grandis* were successfully integrated in a short-season cotton production system, reducing the number of insecticide applications from 10 to 3 with a slight but significant increase in yield (King *et al.*, 1995). In another study, inundative releases of *C. grandis* successfully eliminated reproduction of boll weevils in commercial organic cotton in combination with microbial insecticides and prevented outbreaks of secondary pests by preserving natural enemies in San Angelo, Texas during the 1995 production season (Coleman *et al.*, 1996). The use of inundative releases of *C. grandis* has great potential in the future of boll weevil IPM because of the flexibility in its application. The most important step in the commercialization of this technology was the development of an artificial diet capable of producing high numbers of efficient parasitoids (Rojas *et al.*, 1996; Morales-Ramos *et al.*, 1998). *C. grandis* has been mass-produced on the meridic diet of Rojas *et al.* (1996) in sufficient numbers to successfully control the boll weevil with inundative releases in two 4-hectare experimental cotton fields in the Rio Grande Valley (R.J. Coleman and J.A. Morales-Ramos, unpublished). When published, this will be the first report of an ectoparasitoid of a primary pest in an annual crop reared on a meridic diet being successfully released in the field and achieving control of the target pest.

As discussed above, the main limitation for the use of inundative releases is the high cost of mass propagation and distribution of natural enemies. Factors that may offset this limitation may include multiple insecticide resistance, targeting of key pests of high-value crops, and their use within organically grown crops. One good candidate is the Colorado potato beetle, *Leptinotarsa decemlineata* (Say), which is one of the most important pests of potatoes and tomatoes in North America and Europe (Schalk and Stoner, 1979; Hare, 1980). This important pest has developed resistance to most insecticides used for its control (Grafius, 1995) and it has the potential of developing resistance to *Bacillus thuringiensis* toxins (Whalon *et al.*, 1993). Augmentative releases of the predatory stinkbug, *P. bioculatus*, have been proven effective in reducing populations of *L. decemlineata* and damage to potato crops (Biever and Chauvin, 1992; Hough-Goldstein and Whalen, 1993; Cloutier and Bauduin, 1995). The mass rearing of *P. bioculatus* is costly and complicated (Tamaki and Butt, 1978), but this predator has been successfully reared on an artificial diet for 11 generations (Rojas *et al.*, 2000), with indication of potential to simplify the rearing system. The eulophid egg parasitoid *Edovum puttleri* Grissell is another promising natural enemy of *L. decemlineata* that can inflict 80 percent or higher mortality (Lashomb *et al.*, 1987a,b). Hare (1990) considers this parasitoid to be a good candidate for inundative releases based on its efficacy (>80%) to inflict egg mortality, its host specificity, and its inability to overwinter in the northern USA.

Conclusions

Natural enemies, by regulating pest populations, constitute an important part of the IPM practice in all crops and urban environments. Application of biological control requires the consideration of environmental and biological factors associated with the dynamics of pest populations. Biological control offers three tactics — importation, conservation, and augmentation — that have been proven effective and can be applied individually or in an integrated approach. Primary pests of annual crops are not immune to biological control, nor are biological control and pesticides mutually exclusive strategies. Therefore, there is no justification for excluding biological control from any IPM program. In fact, the application of at least one biological control tactic (conservation) is a requirement for IPM qualification in accordance with the definitions of IPM (Kogan, 1998).

Biological control as a pest management tactic is more successful when applied within the context of IPM. The use of simulation models in agricultural systems has been one of the most important contributions of IPM (Coulson and Saunders, 1987; Gutierrez and Wilson, 1989). Simulation models have made significant contributions to the success of biological control implementation in many crops (Legaspi *et al.*, 1995).

Developments in mass propagation and augmentation biological control will probably make more significant contributions in IPM systems in the future. As the technology of mass propagation of natural enemies develops, more arthropod pest species will become amenable to biological control. The flexibility provided by inundative releases of natural enemies aided by simulation models makes this strategy suitable for IPM programs. A new industry of mass propagation of natural enemies is being born as costs of mass rearing are reduced, making this process commercially competitive.

References

Adjei-Maafo, I.K. and Wilson, L.T. (1983) Factors affecting the relative abundance of arthropods in nectaried and nectariless cotton. *Environ. Entomol.*, **12**, 349–352.

Adkisson, P.L. (1971) Objective uses of pesticides in agriculture. In J.E. Swift (ed.), *Agricultural Chemicals: Harmony or Discord for Food, People, Environment*. Proc. Symp. Univ. Calif. Div. Agr. Sci., Sacramento, CA, pp. 43–51.

Andrewartha, H.G. and Birch, L.C. (1982) Theory of the distribution and abundance of animals. In *The Ecological Web: More on the Distribution and Abundance of Animals*, Chicago University Press, Chicago, IL, pp. 185–211.

Annis, B. and O'Keeffe, L.E. (1987) Influence of pea genotype on parasitization of the pea weevil, *Bruchus pisorum* (Coleoptera: Bruchidae) by *Eupteromalus leguminis* (Hymenoptera: Pteromalidae). *Environ. Entomol.*, **16**, 653–655.

Axtell, R.C. and Rutz, D.A. (1986) Role of parasites and predators as biological fly control agents in poultry production facilities. In R.S. Patterson and D.A. Rutz (eds.), *Biological Control of Muscoid Flies*, Entomological Society of America, Miscellaneous Publications No. 61, College Park, MD, pp. 88–100.

Barbosa, P. and Bentry, B. (1998) The influence of plants on insect parasitoids: implications for conservation biological control. In P. Barbosa (ed.), *Conservation Biological Control*, Academic Press, San Diego, CA, pp. 55–82.

Beirne, B.P. (1962) Trends in applied biological control of insects. *Annu. Rev. Entomol.*, **7**, 387–400.

Biever, K.D. and Chauvin, R.L. (1992) Suppression of the Colorado potato beetle (Coleoptera: Chrysomelidae) with augmentative releases of predaceous stinkbugs (Hemiptera: Pentatomidae). *J. Econ. Entomol.*, **85**, 720–726.

Bracken, G.K. (1965) Effects of dietary components on fecundity of the parasitoid *Exeristes comstockii* (Cress.) (Hymenoptera: Ichneumonidae). *Can. Entomol.*, **97**, 1037–1041.

Brust, G.E. (1994) Natural enemies in straw-mulch reduced Colorado potato beetle populations and damage in potato. *Biol. Cont.*, **4**, 163–169.

Brust, G.E., Stinner, B.R. and McCartney, D.A. (1985) Tillage and soil insecticide effects on predator–black cutworm (Lepidoptera: Noctuidae) interactions in corn agroecosystems. *J. Econ. Entomol.*, **78**, 1389–1392.

Brust, G.E., Stinner, B.R. and McCartney, D.A. (1986) Predator activity and predation in corn agroecosystems. *Environ. Entomol.*, **15**, 1017–1021.

Burke, H.R., Clark, W.E., Cate, J.R. and Fryxell, P.A. (1986) Origin and dispersal of the boll weevil. *Bull. Entomol. Soc. Am.*, **32**, 228–238.

Caltagirone, L.E. (1981) Landmark examples in classical biological control. *Annu. Rev. Entomol.*, **26**, 213–232.

Carpenter, J.E. and Greany, P.D. (1998) Comparative development and performance of artificially-reared versus host-reared *Diapetimorpha introita* (Cresson) (Hymenoptera: Ichneumonidae) wasps. *Biol. Cont.*, **11**, 203–208.

Cate, J.R. (1987) A method of rearing parasitoids of boll weevil without the host plant. *Southwest. Entomol.*, **12**, 211–215.

Cate, J.R. (1990) Biological control of pests and diseases: integrating a diverse heritage. In R.R. Baker and P.E. Dunn (eds.), *New Directions in Biological Control: Alternatives for Suppressing Agricultural Pests and Diseases*, Alan R. Liss, New York, pp. 23–43.

Cate, J.R., Krauter, P.C. and Godfrey, K.E. (1990) Pests of cotton. In D.H. Habeck, F.D. Bennett and J.H. Frank (eds.), *Classical Biological Control in the Southern United States*, Southern Cooperative Series Bulletin No. 355, pp. 17–29.

Chesnut, T.L. and Cross, W.H. (1971) Arthropod parasites of the boll weevil, *Anthonomus grandis*: 2. Comparisons of their importance in the United States over a period of thirty-eight years. *Ann. Entomol. Soc. Am.*, **64**, 550–557.

Cloutier, C. and Bauduin, F. (1995) Biological control of the Colorado potato beetle *Leptinotarsa decemlineata* (Coleoptera: Chrysomelidae) in Quebec by augmentative releases of the two-spotted stinkbug *Perillus bioculatus* (Hemiptera: Pentatomidae). *Can. Entomol.*, **127**, 195–212.

Cohen, A.C. (1981) An artificial diet for *Geocoris punctipes. Southwest. Entomol.*, **6**, 109–113.

Cohen, A.C. (1985) A simple method for rearing the insect predator *Geocoris punctipes* (Heteroptera: Lygaeidae) on a meat diet. *J. Econ. Entomol.*, **78**, 1173–1175.

Cohen, A.C. and Smith, L.K. (1998) A new concept in artificial diets for *Chrysoperla rufilabris*: the efficacy of solid diets. *Biol. Cont.*, **13**, 49–54.

Cohen, A.C. and Staten, R.T. (1994) Long-term culturing and quality assessment of predatory big-eyed bugs, *Geocoris punctipes*. In S.K. Narang, A.C. Bartlett and R.M. Faust (eds.), *Applications of Genetics to Arthropods of Biological Control Significance*, CRC Press, Boca Raton, FL, pp. 121–132.

Coleman, R.J., Morales-Ramos, J.A., King, E.G. and Wood, L.A. (1996) Suppression of boll weevil in organic cotton by release of *Catolaccus grandis* as part of the Southern Rolling Plain boll weevil eradication program. In P. Dugger and D.A. Richer (eds.), *Proc. Beltwide Cotton Conf.*, Cotton Council of America, Memphis, TN, p. 1094.

Corbett, A. and Plant, R.E. (1993) Role of movement in the response of natural enemies to agroecosystem diversification: a theoretical evaluation. *Environ. Entomol.*, **22**, 519–531.

Coulson, R.N. and Saunders, M.C. (1987) Computer-assisted decision-making as applied to entomology. *Annu. Rev. Entomol.*, **32**, 415–437.

Currier, W.W., MacCollom, G.B. and Baumann, G.L. (1982) Drift residues of air applied carbaryl in an orchard environment. *J. Econ. Entomol.*, **75**, 1062–1068.

Curry, G.L., Cate, J.R. and Sharpe, P.J.H. (1982) Cotton bud drying: contributions to boll weevil mortality. *Environ. Entomol.*, **11**, 344–350.

DeBach, P. (ed.) (1964) *Biological Control of Insect Pests and Weeds*, Chapman and Hall, London.

DeBach, P. and Hagen, K.S. (1964) Manipulation of entomophagous species. In P. DeBach (ed.), *Biological Control of Insect Pests and Weeds*, Chapman and Hall, London, pp. 429–458.

DeBach, P. and White, E.B. (1960) Commercial mass culture of the California red scale parasite *Aphytis lingnanensis. Calif. Agric. Exp. Station Bull 770*, 58 pp.

DeBach, P., Rosen, D. and Kennett, C.E. (1971) Biological control of coccids by introduced natural enemies. In C.B. Huffaker (ed.), *Biological Control*, Plenum Press, New York, pp. 165–194.

De Clercq, P. and Degheele, D. (1992) A meat-based diet for rearing the predatory stinkbug *Podisus maculoventris* and *Podisus sagitta. Entomophaga*, **37**, 194–195.

Edwards, R.L. (1954) The effect of diet on egg maturation and resorption in *Mormoniella vitripennis* (Hymenoptera, Pteromalidae). *Quart. J. Micr. Sci.*, **95**, 459–468.

Edwards, R.H., Morales-Ramos, J.A., Rojas, M.G. and King, E.G. (1998) Proposed transformation of laboratory to commercial-scale mass rearing of *Catolaccus grandis* (Hymenoptera: Pteromalidae). *Vedalia*, **5**, 133–139.

Ehler, L.E. (1998) Conservation biological control: past, present, and future. In P. Barbosa (ed.), *Conservation Biological Control*, Academic Press, San Diego, CA, pp. 1–8.

Ehler, L.E. and Endicott, P.C. (1984) Effect of malathion-bait sprays on biological control of insect pests of olive, citrus and walnut. *Hilgardia*, **52**, 1–47.

Elzen, G.W., Williams, H.J. and Vinson, S.B. (1983) Response by the parasitoid *Campoletis sonorensis* (Hymenoptera: Ichneumonidae) to chemicals (synomones) in plants: implications for host habitat location. *Environ. Entomol.*, **12**, 1873–1877.

Elzen, G.W. and King, E.G. (1999) Periodic releases and manipulation of natural enemies. In T.S. Bellows and T.W. Fisher (eds.), *Handbook of Biological Control*, Academic Press, San Diego, CA, pp. 253–270.

Fillman, D.A. and Sterling, W.L. (1983) Killing power of the red imported fire ant (Hym.: Formicidae): a key predator of the boll weevil (Col.: Curculionidae). *Entomophaga*, **28**, 339–344.

Fillman, D.A. and Sterling, W.L. (1985) Inaction levels for the red imported fire ant, *Solenopsis invicta* (Hym.: Formicidae): a predator of the boll weevil, *Anthonomus grandis* (Col.: Curculionidae). *Agric. Ecosyst. Environ.*, **13**, 93–102.

Flanders, R.V. (1985) Biological control of the Mexican bean beetle: potentials for and problems of inoculative releases of *Pediobius foveolatus*. In R. Shibles (ed.), *World Soybean Research Conference III: Proceedings*, Westview, Boulder, CO.

Flanders, S.E. (1935) An apparent correlation between the feeding habits of certain pteromalids and the condition of their ovarian follicles. *Ann. Entomol. Soc. Am.*, **28**, 438–444.

Flanders, K.L. and Radcliffe, E.B. (1999) Alfalfa IPM. In E.B. Radcliffe and W.D. Hutchison (eds.), *Radcliffe's IPM World Textbook*, URL: http://ipmworld.umn.edu, University of Minnesota, St. Paul, MN.

Frisbie, R.E. and Adkisson, P.L. (1985) IPM: definitions and current status in U.S. agriculture. In M.A. Hoy and D.C. Herzog (eds.), *Biological Control in Agricultural IPM Systems*, Academic Press, Orlando, FL, pp. 41–51.

Frisbie, R.E., Reynolds, H.T., Adkisson, P.L. and Smith, R.F. (1994) Cotton insect pest management. In R.L. Metcalf and W.H. Luckmann (eds.), *Introduction to Insect Pest Management*, 3rd edn, Wiley, New York, pp. 421–468.

Furlong, M.J., Verkerk, H.J. and Wright, D.J. (1994) Differential effects of the acylurea insect growth regulator teflubenzuron on the adults of two endolarval parasitoids of *Plutella xylostella, Cotesia plutellae* and *Diadegma semiclausum*. *Pestic. Sci.*, **41**, 359–364.

Grafius, E.J. (1995) Is local selection followed by dispersal a mechanism for rapid development of multiple insecticide resistance in the Colorado potato beetle? *Am. Entomol.*, **41**, 104–109.

Greany, P.D. and Carpenter, J.E. (1998) Culture medium for parasitic and predaceous insects. *U.S. Patent* 5,799,607.

Greathead, D.J. (1986) Parasitoids in classical biological control. In J. Waage and D. Greathead (eds.), *Insect Parasitoids*, Academic Press, Orlando, FL, pp. 289–318.

Gutierrez, A.P. and Wilson, L.T. (1989) Development and use of pest models. In R.E. Frisbie, K.M. El-Zik and L.T. Wilson (eds.), *Integrated Pest Management Systems and Cotton Production*, Wiley, New York, pp. 65–83.

Hagen, K.S., Sawall, E.F. Jr. and Tassan, R.L. (1971) The use of food sprays to increase effectiveness of entomophagous insects. *Proc. 2nd Tall Timbers Conference on Ecological Animal Control by Habitat Management*, Tall Timbers Research Station, Tallahassee, FL, pp. 59–81.

Hare, J.D. (1980) Impact of defoliation by the Colorado potato beetle on potato yields. *J. Econ. Entomol.*, **73**, 369–373.

Hare, J.D. (1990) Ecology and management of the Colorado potato beetle. *Annu. Rev. Entomol.*, **35**, 81–100.

Hasagawa, M.K., Niijima, K. and Matsuka, M. (1989) Rearing *Chrysoperla carnea* (Neuroptera: Chrysopidae) on chemically defined diets. *Appl. Entomol. Zool.*, **24**, 96–102.

Hoddle, M., Van Driesche, R. and Sanderson, J. (1997) Biological control of *Bemisia argentifolii* (Homoptera: Aleyrodidae) on poinsettia with inundative releases of *Encarsia formosa* (Hymenoptera: Aphelinidae): are higher release rates necessarily better? *Biol. Cont.*, **10**, 166–179.

Hoddle, M.S., Van Driesche, R.G., Sanderson, J.P. and Minkenberg, P.P.J.M. (1998) Biological control of *Bemisia argentifolii* (Homoptera: Aleyrodidae) on poinsettia with inundative releases of *Eretmocerus eremicus* (Hymenoptera: Aphelinidae): do release rates affect parasitism? *Bull. Entomol. Res.*, **88**, 47–58.

Hoddle, M.S., Sanderson, J.P. and Van Driesche, R.G. (1999) Biological control of *Bemisia argentifolii* (Homoptera: Aleyrodidae) on poinsettia with inundative releases of *Eretmocerus eremicus* (Hymenoptera: Aphelinidae): does varying the weekly release rate affect control? *Bull. Entomol. Res.*, **89**, 41–51.

Horn, D.J. (1988) *Ecological Approach to Pest Management*. The Guilford Press, New York.

Hough-Goldstein, J. and Whalen, J. (1993) Inundative release of predatory stink bugs for control of Colorado potato beetle. *Biol. Cont.*, **3**, 343–347.

House, H.L. (1977) Nutrition of natural enemies. In R.L. Ridway and S.B. Vinson (eds.), *Biological Control by Augmentation of Natural Enemies*, Plenum Press, New York, pp. 151–182.

Hoy, M.A. (1985) Improving establishment of arthropod natural enemies. In M.A. Hoy and D.C. Herzog (eds.), *Biological Control in Agricultural IPM Systems*, Academic Press, Orlando, FL, pp. 151–166.

Hoy, M.A. (1994) Parasitoids and predators in management of arthropod pests. In R.L. Metcalf and W.H. Luckmann (eds.), *Introduction to Insect Pest Management*, 3rd edn, Wiley, New York, pp. 129–198.

Huffaker, C.B., Messenger, P.S. and DeBach, P. (1971) The natural enemy component in natural control and the theory of biological control. In C.B. Huffaker (ed.), *Biological Control*, Plenum Press, New York, pp. 16–67.

Hull, L.A. and Beers, E.H. (1985) Ecological selectivity: modifying chemical control practices to preserve natural enemies. In M.A. Hoy and D.C. Herzog (eds.), *Biological Control in Agricultural IPM Systems*, Academic Press, Orlando, FL, pp. 103–122.

Hunter, C.D. (1997) *Suppliers of Beneficial Organisms in North America*, California Environmental Protection Agency, Department of Pesticide Regulation, Sacramento, CA.

Idoine, K. and Ferro, D.N. (1988) Aphid honeydew as a carbohydrate source for *Edovum puttleri* (Hymenoptera: Eulophidae). *Environ. Entomol.*, **17**, 941–944.

Idris, A.B. and Grafius, E. (1996) Effects of wild and cultivated host plants on oviposition, survival, and development of diamondback moth (Lepidoptera: Plutellidae) and its parasitoid *Diadegma insulare* (Hymenoptera: Ichneumonidae). *Environ. Entomol.*, **25**, 825–833.

Johnson, M.T. and Gould, F. (1992) Interaction of genetically engineered host plant resistance and natural enemies of *Heliothis virescens* (Lepidoptera: Noctuidae) in tobacco. *Environ. Entomol.*, **21**, 586–597.

Johnson, W.L., Cross, W.H., McGovern, W.L. and Mitchell, H.C. (1973) Biology of *Heterolaccus grandis* in a laboratory culture and its potential as an introduced parasite of the boll weevil in the United States. *Environ. Entomol.*, **2**, 112–118.

King, E.G. (1993) Augmentation of parasites and predators for suppression of arthropod pests. In R.D. Lumsden and J.L. Vaughn (eds.), *Pest Management: Biologically Based Technologies*, Conference Proceeding Series, American Chemical Society, Washington, DC, pp. 90–100.

King, E.G. and Morrison, R.K. (1984) Some systems for production of eight entomophagous arthropods. In E.G. King and N.C. Leppla (eds.), *Advances and Challenges in Insect Rearing*, U.S. Department of Agriculture, Agricultural Research Service, New Orleans, LA, pp. 206–222.

King, E.G. and Powell, J.E. (1992) Propagation and release of natural enemies for control of cotton insect and mite pests in the United States. *Crop Prot.*, **11**, 497–506.

King, E.G., Hopper, K.R. and Powell, J.E. (1985) Analysis of systems for biological control of crop arthropod pests in the U.S. by augmentation of predators and parasites. In M.A. Hoy and D.C. Herzog (eds.), *Biological Control in Agricultural IPM Systems*, Academic Press, Orlando, FL, pp. 201–227.

King, E.G., Summy, K.R., Morales-Ramos, J.A. and Coleman, R.J. (1993) Integration of boll weevil biological control by inoculative/augmentative releases of the parasite *Catolaccus grandis* in short season cotton. In D.J. Herber and D.A. Richter (eds.), *Proc. Beltwide Cotton Conf.*, Cotton Council of America, Memphis, TN, pp. 910–914.

King, E.G., Coleman, R.J., Wood, L., Wendel, L., Greenberg, S. and Scott, A.W. (1995) Suppression of the boll weevil in commercial cotton by augmentative releases of a wasp parasite, *Catolaccus grandis*. In D.A. Richter and J. Armour (eds.), *Addendum to the Proc. Beltwide Cotton Conf.*, Cotton Council of America, Memphis, TN, pp. 26–30.

King, E.G., Coleman, R.J., Morales-Ramos, J.A. and Summy, K.R. (1996) Biological control. In E.G. King, J.R. Phillips and R.J. Coleman (eds.), *Cotton Insects and Mites: Characterization and Management*, The Cotton Foundation Reference Book Series, No. 3, The Cotton Foundation Publisher, Memphis, TN, pp. 511–538.

Kingsley, P.C., Bryan, M.D., Day, W.H., Burger, T.L., Dysart, R.J. and Schwalbe, C.P. (1993) Alfalfa weevil (Coleoptera: Curculionidae) biological control: spreading the benefits. *Environ. Entomol.*, **22**, 1234–1250.

Knipling, E.F. (1977) The theoretical basis for augmentation of natural enemies. In R.L. Ridgway and S.B. Vinson (eds.), *Biological Control by Augmentation of Natural Enemies*, Plenum Press, New York, pp. 79–123.

Kogan, M. (1998) Integrated pest management: historical perspectives and contemporary developments. *Annu. Rev. Entomol.*, **43**, 243–270.

Kogan, M. and Lattin, J.D. (1993) Insect conservation and pest management. *Biodiv. Conserv.*, **2**, 242–257.

Kogan, M. and Turnipseed, S.G. (1987) Ecology and management of soybean arthropods. *Annu. Rev. Entomol.*, **32**, 507–538.

Kogan, M., Gerling, D. and Maddox, J.V. (1999) Enhancement of biological control in annual agricultural environments. In T.S. Bellows and T.W. Fisher (eds.), *Handbook of Biological Control*, Academic Press, San Diego, CA, pp. 789–818.

Juan A. Morales-Ramos and M. Guadalupe Rojas

Landis, D.A. and Haas, M.J. (1992) Influence of landscape structure on abundance and within field distribution of European corn borer (Lepidoptera: Pyralidae) larval parasitoids in Michigan. *Environ. Entomol.*, **21**, 409–416.

Lashomb, J., Krainacker, D., Jansson, R.K., Ng, Y.S. and Chianese, R. (1987a) Parasitism of *Leptinotarsa decemlineata* (Say) eggs by *Edovum puttleri* Griessell (Hymenoptera: Eulophidae): effects of host age, parasitoid age, and temperature. *Can. Entomol.*, **119**, 75–82.

Lashomb, J., Ng, Y.S., Jansson, R.K. and Bullock, R. (1987b) *Edovum puttleri* (Hymenoptera: Eulophidae), an egg parasitoid of Colorado potato beetle (Coleoptera: Chrysomelidae): development and parasitism on eggplant. *J. Econ. Entomol.*, **80**, 65–68.

Legaspi, B.C. Jr., Carruthers, R.I. and Morales-Ramos, J.A. (1995) Simulation modelling of biological control systems. *Vedalia*, **2**, 43–60.

Leius, K. (1960) Attractiveness of different foods and flowers to the adults of some hymenopterous parasites. *Can. Entomol.*, **92**, 369–376.

Leius, K. (1961) Influence of various foods on fecundity and longevity of adults of *Scambus buolianae* (Htg.) (Hymenoptera: Ichneumonidae). *Can. Entomol.*, **93**, 1079–1084.

Lingren, P.D. and Lukefahr, M.J. (1977) Effects of nectariless cotton on caged populations of *Campoletis sonorensis*. *Environ. Entomol.*, **6**, 586–588.

Luck, R.F. (1981) Parasitic insects introduced as biological control agents for arthropod pests. In D. Pimentel (ed.), *CRC Handbook of Insect Pest Management in Agriculture*, vol. II, CRC Press, Boca Raton, FL, pp. 125–284.

Luckmann, W.H. and Metcalf, R.L. (1994) The pest-management concept. In R.L. Metcalf and W.H. Luckmann (eds.), *Introduction to Insect Pest Management*, 3rd edn, Wiley, New York, pp. 1–34.

Lum, P.T.M. (1977) Effects of glucose on autogenous reproduction of *Bracon hebetor* Say. *J. Georgia Entomol. Soc.*, **12**, 150–153.

Messenger, P.S. and van den Bosch, R. (1971) The adaptability of introduced biological control agents. In C.B. Huffaker (ed.), *Biological Control*, Plenum Press, New York, NY, pp. 68–92.

Metcalf, R.L. (1994) Insecticides in pest management. In R.L. Metcalf and W.H. Luckmann (eds.), *Introduction to Insect Pest Management*, 3rd edn, Wiley, New York, pp. 245–314.

Meyer, J.A. (1986) Biological control of filth flies associated with confined livestock. In R.S. Patterson and D.A. Rutz (eds.), *Biological Control of Muscoid Flies*, Entomological Society of America, Miscellaneous Publications No. 61, College Park, MD, pp. 108–115.

Morales-Ramos, J.A. and Cate, J.R. (1992) Rate of increase and adult longevity of *Catolaccus grandis* (Burks) (Hymenoptera: Pteromalidae) in the laboratory at four temperatures. *Environ. Entomol.*, **21**, 620–627.

Morales-Ramos, J.A. and King, E.G. (1991) Evaluation of *Catolaccus grandis* (Burks) as a biological control agent against the cotton boll weevil. In D.J. Herber and D.A. Richter (eds.), *Proc. Beltwide Cotton Conf.*, Cotton Council of America, Memphis, TN, p. 724.

Morales-Ramos, J.A., Summy, K.R., Roberson, J.L., Cate, J.R. and King, E.G. (1992) Feasibility of mass rearing *Catolaccus grandis*, a parasitoid of the boll weevil. In D.J. Herber and D.A. Richter (eds.), *Proc. Beltwide Cotton Conf.*, Cotton Council of America, Memphis, TN, pp. 723–726.

Morales-Ramos, J.A., Rojas, M.G., Roberson, R.G., King, E.G., Summy, K.R. and Brazzel, J.R. (1994) Suppression of the boll weevil first generation by augmentative releases of *Catolaccus grandis* in Aliceville, Alabama. In D.J. Herber and D.A. Richter (eds.), *Proc. Beltwide Cotton Conf.*, Cotton Council of America, Memphis, TN, pp. 958–964.

Morales-Ramos, J.A., Summy, K.R. and King, E.G. (1995) Estimating parasitism by *Catolaccus grandis* (Hymenoptera: Pteromalidae) after inundative releases against the boll weevil (Coleoptera: Curculionidae). *Environ. Entomol.*, **24**, 1718–1725.

Morales-Ramos, J.A., Summy, K.R. and King, E.G. (1996) ARCASIM, a model to evaluate augmentation strategies of the parasitoid *Catolaccus grandis* against boll weevil populations. *Ecol. Modelling*, **93**, 221–235.

Morales-Ramos, J.A., Rojas, M.G., Coleman, R.J. and King, E.G. (1998) Potential use of *in vitro*-reared *Catolaccus grandis* (Hymenoptera: Pteromalidae) for biological control of the boll weevil (Coleoptera: Curculionidae). *J. Econ. Entomol.*, **91**, 101–109.

Morgan, P.B. (1986) Mass culturing microhymenopteran pupal parasites (Hymenoptera: Pteromalidae) of filth breeding flies. In R.S. Patterson and D.A. Rutz (eds.), *Biological Control of Muscoid Flies*, Entomological Society of America, Miscellaneous Publications No. 61, College Park, MD, pp. 77–87.

Morrison, D.E., Bradley, J.R. Jr. and Van Duyn, J.W. (1979) Populations of corn earworm and associated predators after applications of certain soil-applied pesticides to soybeans. *J. Econ. Entomol.*, **72**, 97–100.

Murchie, A.K., Williams, I.H. and Alford D.V. (1997) Effects of commercial insecticide treatments to winter oilseed rape on parasitism of *Ceutorhynchus assimilis* Paykull (Coleoptera: Curculionidae) by *Trichomalus perfectus* (Walker) (Hymenoptera: Pteromalidae). *Crop Prot.*, **16**, 199–202.

Murdock, W.W. and Oaten, A. (1975) Predation and population stability. *Adv. Ecol. Res.*, **9**, 1–131.

Murphy, B.C., Rosenheim, J.A. and Granett, J. (1996) Habitat diversification for improving biological control: abundance of *Anagrus epos* (Hymenoptera: Mymaridae) in grape vineyards. *Environ. Entomol.*, **25**, 495–504.

National Academy of Sciences (1969) *Insect Pest Management and Control*, Publ. 1695. Washington, DC.

Pair, S.D., Laster, M.L. and Martin, D.F. (1982) Parasitoids of *Heliothis* spp. (Lepidoptera: Noctuidae) larvae in Mississippi associated with sesame interplantings in cotton, 1971–1974: implications of host–habitat interaction. *Environ. Entomol.*, 11, 509–512.

Parrella, M.P., Heinz, K.M. and Nunney, L. (1992) Biological control through augmentative releases of natural enemies: a strategy whose time has come. *Am. Entomol.*, **38**, 172–179.

Patterson, R.S. and Morgan, P.B. (1986) Factors affecting the use of an IPM scheme at poultry installations in a semi-tropical climate. In R.S. Patterson and D.A. Rutz (eds.), *Biological Control of Muscoid Flies*, Entomological Society of America, Miscellaneous Publications No. 61, College Park, MD, pp. 101–107.

Powell, W. (1986) Enhancing parasitoid activity in crops. In J. Waage and D. Greathead (eds.), *Insect Parasitoids*, Academic Press, Orlando, FL, pp. 319–340.

Price, P.W. (1984) *Insect Ecology*, 2nd edn, Wiley Interscience, New York.

Price, P.W. (1987) The role of natural enemies in insect populations. In P. Barbosa and J.C. Schultz (eds.), *Insect Outbreaks*, Academic Press, San Diego, CA, pp. 287–312.

Price, P.W. (1992) Three-trophic-level interactions affecting the success of biological control projects. *Pesq. Agropec. Bras.*, **27**, 15–29.

Price, P.W. and Waldbauer, G.P. (1994) Ecological aspects of pest management. In R.L. Metcalf and W.H. Luckmann (eds.), *Introduction to Insect Pest Management*, 3rd edn, Wiley, New York, pp. 35–65.

Quicke, D.L.J. (1997) *Parasitic Wasps*. Chapman and Hall, London.

Rabb, R.L., Stinner, R.E. and van den Bosch, R. (1976) Conservation and augmentation of natural enemies. In C.B. Huffaker and P.S. Messenger (eds.), *Theory and Practice of Biological Control*, Academic Press, New York, pp. 233–254.

Radcliffe, E.B. and Flanders, K.L. (1998) Biological control of alfalfa weevil in North America. *Integ. Pest Manage. Rev.*, **3**, 225–242.

Riggin, T.M., Wiseman, B.R., Isenhour, D.J. and Espelie, K.E. (1992) Incidence of fall armyworm (Lepidoptera: Noctuidae) parasitoids on resistant and susceptible corn genotypes. *Environ. Entomol.*, **21**, 888–895.

Risch, S.J. (1987) Agricultural ecology and insect outbreaks. In P. Barbosa and J.C. Schultz (eds.), *Insect Outbreaks*, Academic Press, San Diego, CA, pp. 217–238.

Roberson, J.L. and Wright, J.E. (1984) Production of boll weevils, *Anthonomus grandis*. In E.G. King and N.C. Leppla (eds.), *Advances and Challenges in Insect Rearing*, USDA, ARS, New Orleans, LA, pp. 188–192.

Rojas, M.G., Morales-Ramos, J.A. and King, E.G. (1996) *In vitro* rearing of the boll weevil (Coleoptera: Curculionidae) ectoparasitoid *Catolaccus grandis* (Hymenoptera: Pteromalidae) on meridic diets. *J. Econ. Entomol.*, **89**, 1095–1104.

Rojas, M.G., Morales-Ramos, J.A. and King, E.G. (2000) Two meridic diets for *Perillus bioculatus* (Heteroptera: Pentatomidae), a predator of *Leptinotarsa decemlineata* (Coleoptera: Chrysomelidae). *Biol. Cont.*, **17**, 92–99.

Rojas, M.G., Morales-Ramos, J.A. and King, E.G. (1999) Synthetic diet for rearing the hymenopterous ectoparasitoid, *Catolaccus grandis*. *U.S. Patent* 5,899,168.

Ruberson, J.R., Herzog, G.A., Lambert, W.R. and Lewis, W.J. (1994) Management of the beet armyworm (Lepidoptera: Noctuidae) in cotton: role of natural enemies. *Fla. Entomol.*, **77**, 440–453.

Ruberson, J.R., Nemoto, H. and Hirose, Y. (1998) Pesticides and conservation of natural enemies in pest management. In P. Barbosa (ed.), *Conservation Biological Control*, Academic Press, San Diego, CA, pp. 207–220.

Schalk, J.M. and Stoner, A.K. (1979) Tomato production in Maryland: Effects of different densities of larvae and adults of the Colorado potato beetle. *J. Econ. Entomol.*, **72**, 826–829.

Shepard, M., Carner, G.R. and Turnipseed, S.G. (1977) Colonization and resurgence of insect pests of soybean in response to insecticides and field isolation. *Environ. Entomol.*, **6**, 501–506.

Smith, S.M. (1996) Biological control with *Trichogramma*: advances, successes, and potential of their use. *Annu. Rev. Entomol.*, **41**, 375–406.

Steffey, K.L., Armbrust, E.J. and Onstad, D.W. (1994) Management of insects in alfalfa. R.L. Metcalf and W.H. Luckmann (eds.), *Introduction to Insect Pest Management*, 3rd edn, Wiley, New York, pp. 469–506.

Sterling, W.L., El-Zik, K.M. and Wilson, L.T. (1989) Biological control of pest populations. In R.E. Frisbie, K.M. El-Zik and L.T. Wilson (eds.), *Integrated Pest Management Systems and Cotton Production*, Wiley, New York, pp. 155–189.

Juan A. Morales-Ramos and M. Guadalupe Rojas

Stern, V.M., Smith, R.F., van den Bosch, R. and Hagen, K.S. (1959) The integrated control concept. *Hilgardia*, **29**, 81–101.

Stinner, R.E. (1977) Efficacy of inundative releases. *Annu. Rev. Entomol.*, **22**, 515–531.

Stinner, B.R. and House, G.J. (1990) Arthropods and other invertebrates in conservation tillage agriculture. *Annu. Rev. Entomol.*, **35**, 299–318.

Streams, F.A. and Fuester, R.W. (1967) Biology and distribution of *Tetrastichus incertus*, a parasite of the alfalfa weevil. *J. Econ. Entomol.*, **60**, 1574–1579.

Sturm, M.M. and Sterling, W.L. (1986) Boll weevil mortality factors within flower buds of cotton. *Bull. Entomol. Soc. Am.*, **32**, 239–247.

Sturm, M.M., Sterling, W.L. and Dean, D.A. (1989) Life table and key mortality factors of boll weevils in Texas. *Texas Agric. Exp. Sta. Misc. Publ.* MP-1675, College Station, TX.

Sturm, M.M., Sterling, W.L. and Hartstack, A.W. (1990) Role of natural mortality in boll weevil (Coleoptera: Curculionidae) management programs. *J. Econ. Entomol.*, **83**, 1–7.

Summy, K.R., Morales-Ramos, J.A., King, E.G., Coleman, R.J. and Scott, A.W. (1994a) Impact of mechanical cultivation on searching efficiency of *Catolaccus grandis*, and exotic parasite of boll weevil. *Southwest Entomol.*, **19**, 379–384.

Summy, K.R., Morales-Ramos, J.A., King, E.G., Wolfenbarger, D.A., Coleman, R.J. and Greenberg, S.M. (1994b) Integration of boll weevil parasite augmentation into the short-season cotton production system of the lower Rio Grande Valley. In D.J. Herber and D.A. Richter (eds.), *Proc. Beltwide Cotton Conf.*, Cotton Council of America, Memphis, TN, pp. 953–957.

Summy, K.R., Morales-Ramos, J.A. and King, E.G. (1995) Suppression of boll weevil (Coleoptera: Curculionidae) infestations on South Texas cotton by augmentative releases of the parasite *Catolaccus grandis* (Hymenoptera: Pteromalidae). *Biol. Cont.*, **5**, 523–529.

Summy, K.R., Raulston, J.R., Spurgeon, D. and Vargas, J. (1996) An analysis of the beet armyworm outbreak on cotton in the lower Rio Grande Valley of Texas during the 1995 production season. In P. Dugger and D.A. Richer (eds.), *Proc. Beltwide Cotton Conf.*, Cotton Council of America, Memphis, TN, pp. 837–842.

Summy, K.R., Raulston, J.R., Spurgeon, D.W. and Scott, A.W. Jr. (1997) Population trends of beet armyworm on cotton in the lower Rio Grande Valley during the 1996 production season. In P. Dugger and D.A. Richer (eds.), *Proc. Beltwide Cotton Conf.*, Cotton Council of America, Memphis, TN, pp. 1035–1039.

Syme, P.D. (1975) The effect of flowers on the longevity and fecundity of two native parasites of European pine shoot moth in Ontario. *Environ. Entomol.*, **4**, 337–346.

Tamaki, G. and Butt, B.A. (1978) Impact of *Perillus bioculatus* on the Colorado potato beetle and plant damage. *USDA Technical Bull. No. 1581*, Washington, DC.

Thompson, S.N. (1986) Nutrition and *in vitro* culture of insect parasitoids. *Annu. Rev. Entomol.*, **31**, 197–219.

Thompson, S.N. (1999) Nutrition and culture of entomophagous insects. *Annu. Rev. Entomol.*, **44**, 561–592.

Thompson, S.N. and Hagen, K.S. (1999) Nutrition of entomophagous insects and other arthropods. In S. Bellows and T.W. Fisher (eds.), *Handbook of Biological Control*, Academic Press, San Diego, CA, pp. 594–652.

Van Driesche, R.G. and Bellows, T.S. Jr. (1996) *Biological Control*, Chapman and Hall, New York, NY.

Van Lenteren, J.C., van Roermund, H.J.W. and Sütterlin, S. (1996) Biological control of greenhouse whitefly (*Trialeurodes vaporariorum*) with the parasitoid *Encarsia formosa*: how does it work? *Biol. Cont.*, **6**, 1–10.

Verkerk, R.H.J., Leather, S.R. and Wright, D.J. (1998) The potential for manipulating crop–pest–natural enemy interactions for improved insect pest management. *Bull. Entomol. Res.*, **88**, 493–501.

Waller, W.E. (1987) Factors affecting insect population dynamics: differences between outbreak and non-outbreak species. *Annu. Rev. Entomol.*, **32**, 317–340.

Whalon, M.E., Miller, D.L., Hollingworth, R.M., Grafius, E.J. and Miller, J.R. (1993) Selection of a Colorado potato beetle (Coleoptera: Chrysomelidae) strain resistant to *Bacillus thuringiensis*. *J. Econ. Entomol.*, **86**, 226–233.

White, A.J., Wratten, S.D., Berry, N.A. and Weigmann, U. (1995) Habitat manipulation to enhance biological control of *Brassica* pests by hover flies (Diptera: Syrphidae). *J. Econ. Entomol.*, **88**, 1171–1176.

Wilson, M.C. and Armbrust, E.J. (1970) Approach to integrated control of the alfalfa weevil. *J. Econ. Entomol.*, **63**, 554–557.

Wilson, L.J., Bauer, L.R. and Lally, D.A. (1998) Effect of early season insecticide use on predators and outbreaks of spider mites (Acari: Tetranychidae) in cotton. *Bull. Entomol. Res.*, **88**, 477–488.

Zchori-Fein, E., Roush, R.T. and Sanderson, J.P. (1994) Potential for integration of biological and chemical control of greenhouse whitefly (Homoptera: Aleyrodidae) using *Encarsia formosa* (Hymenoptera: Aphelinidae) and abamectin. *Environ. Entomol.*, **23**, 1277–1282.

Biological Control by Augmentation of Natural Enemies: Retrospect and Prospect

3

Gary W. Elzen[1], Randy J. Coleman[1], and Edgar G. King[2]
[1]*Kika de la Garza Subtropical Agricultural Research Center, Weslaco, USA,*
[2]*US Department of Agriculture, Agricultural Research Service, Stoneville, USA*

Introduction

Suppression of insect and mite pests by entomophagous arthropods, parasitoids, and predators is well known (DeBach, 1964; Hagen *et al.*, 1971; Huffaker and Messenger, 1976). However, these entomophages may fail to maintain pests below acceptable population levels. Furthermore, the development of insecticide-resistant populations of pests, evidence for resurgence on a worldwide basis (Waage, 1989), and environmental concerns have led to increased research in the area of alternative strategies to pesticides for control of damaging pest populations. Exotic parasitoids and predators may be introduced for establishment (classical biological control). Classical biological control and the successes in that area provide background and encouragement for increased efforts in increasing entomophage numbers within a defined area (biological control by augmentation) (Huffaker, 1974; Greathead, 1986). Methods for manipulation of natural enemies for the purposes of reducing herbivore populations and alleviating plant stress include, but are not limited to, conservation and augmentation, habitat management (Letourneau and Altieri, 1999), and genetic manipulation (Whitten and Hoy, 1999). Herein, we examine some important examples of augmentation programs and explore the avenues available for augmentation of natural enemies to control pests in the context of a rational management program that explores the appropriate use of all available insect control methods.

Augmentation

Augmentation of natural enemies to achieve suppression of pest populations has been reviewed by Beglyarov and Smetnik (1977), Huffaker *et al.* (1977), Knipling (1977), Ridgway and Vinson (1977), Shumakov (1977), Starter and Ridgway (1977), King *et al.* (1984, 1985a), and van Lenteren (1986). Behavior of released parasitoids and predators (Weseloh, 1984) and their effectiveness has also been studied (Vinson, 1975, 1977; Nordlund *et al.*, 1981a,b, 1985; Lewis and Nordlund, 1985).

DeBach and Hagen (1964) described the concepts of inundative and inoculative releases of entomophages. Inundative releases rely mainly on the agents released, not their progeny, whereas inoculative releases rely upon an increase of the initial natural enemy populations that will suppress subsequent pest generations. Inoculative releases may be timed so that both immediate control and subsequent progeny production contribute to pest suppression (Li, 1984). Augmentative or inundative releases have been described by such terms as supplemental releases, strategic releases, programmed releases, seasonal colonization, periodic colonization, and compensatory releases (Ridgway *et al.*, 1977; King *et al.*, 1984, 1985a). Examples demonstrating the feasibility of controlling pests by augmentative releases of entomophages are given in Table 3.1. The major proportions of successfully introduced natural enemies are parasitoids (DeBach, 1964; Laing and Hamai, 1976; van Lenteren, 1986). Greathead (1986) noted that parasitoids have been established and achieved satisfactory control about three times more often than predators. Case studies were presented by King *et al.* (1985a).

Trichogramma

Control of lepidopterous pests by mass rearing and release of *Trichogramma* spp. has been practiced for many decades. The pioneering research of Howard and Fiske (1911) and Flanders (1929, 1930) stimulated research with *Trichogramma* spp., and a number of successes in reducing insect populations by augmentation with *Trichogramma* have been reported. Reviews on the use of *Trichogramma* by Ridgway and Vinson (1977) and Ridgway *et al.* (1977) for the Western Hemisphere, by Huffaker (1977) for China, by Beglyarov and Smetnik (1977) for the former USSR, and by King *et al.* (1985b) for augmentation in cotton are available. As of 1985, *Trichogramma* spp. were the entomophagous arthropods most widely used in augmentation in the world (King *et al.*, 1985b). Hassan (1982) reported 65–93 percent

Table 3.1 Parasitoid and predator species in the USA for which feasibility has been shown in augmentation

Entomophage	Target pest	Commodity	References
Amblyseius californicus (McGregor)	Two-spotted spider mite	Strawberry	Oatman *et al.* (1977)
Anaphes iole (Girault)	*Lygus hesperus* Knight	Strawberry	Norton and Welter (1996)
Aphidius smithi Sharma & Subba Rao	Pea aphid	Pea	Halfhill and Featherston (1973)
Aphytis melinus DeBach	California red scale	Citrus	DeBach and Hagen (1964)
Bracon spp.	Pink bollworm	Cotton	Bryan *et al.* (1973)
			Legner and Medved (1979)
Archytas marmoratus Townsend	*Helicoverpa zea* (Boddie)	Corn	Gross (1990)
Agathis diversa Muesebeck	Oriental fruit moth	Peach	Brunson and Allen (1944)
Chelonus spp.	Pink bollworm	Cotton	Bryan *et al.* (1973)
			Legner and Medved (1979)
Chrysopa carnea Stephens	Several	Several crops	Ables and Ridgway (1981)
Coccinella septempunctata L.	Potato aphid, Green peach aphid	Potato	Shands *et al.* (1972)
Cryptolaemus montrouzieri Mulsant	Citrus mealybug	Citrus	DeBach and Hagen (1964)
Delphastus pusillus (LeConte)	*Bemesia argentifolii* Bellows & Perring	Cotton	Heinz *et al.* (1994)
Diaeretiella rapae McIntosh	Aphids	Potato	Shands *et al.* (1975)
Diglyphus begini (Ashmead)	*Liriomyza trifolii* (Burgess)	Marigolds	Heinz and Parrella (1990)
Encarsia formosa Gahan	Greenhouse whitefly	Greenhouse crops	Vet *et al.* (1980)
E. formosa	*B. argentifolii*	Poinsettia	Hoddle *et al.* (1997)
E. formosa	Greenhouse whitefly	Poinsettia	Zchori-Fein *et al.* (1994)
Encarsia luteola Howard	*B. argentifolii*	Poinsettia	Heinz and Parrella (1994)
Eretmocerus nr. *californicus*	*B. argentifolii*	Cotton	Simmons and Minkenberg (1994)
Euseius tularensis Congdon	Citrus thrips	Citrus	Grafton-Cardwell and Ouyang (1955)
Hippodamia convergens Guérin-Méneville	Aphids	Pea	DeBach and Hagen (1964),
			Cooke (1963)
H. convergens	Melon aphid	Chrysanthemum	Dreistadt and Flint (1996)
Leptomastix dactylopii Howard	Mealybugs	Citrus	Fisher (1963)
Lixophaga diatraeae (Townsend)	Sugarcane borer	Sugarcane	King *et al.* (1982)
Jalysus spinosus (Say)	Tobacco budworm	Tobacco	Ridgway *et al.* (1977)
Aphid parasites	Greenbug, pea aphid	Sorghum, peas	Ridgway *et al.* (1977)
Lysiphlebus testaceipes (Cresson)	Greenbug	Sorghum	Starks *et al.* (1976)
Macrocentrus ancylivorus Rohwer	Oriental fruit moth	Peach	Brunson and Allen (1944)
Microterys flavus (Howard)	Brown soft scale	Citrus	Hart (1972)
Phytoseiulus persimilis Athias-Henriot	Two-spotted spider mite	Strawberry	Oatman *et al.* (1968, 1976, 1977)
	Tobacco budworm	Tobacco	Lawson *et al.* (1961)
Polistes spp.	Lepidopterous larvae	Cotton	Fye (1972)

Table 3.1 *(cont'd)*

Entomophage	Target pest	Commodity	References
Pediobius foveolatus (Crawford)	Mexican bean beetle	Soybean	Ridgway et al. (1977)
Predaceous mites	Several mite pests	Fruit, greenhouse crops	Ridgway et al. (1977)
			Chant (1959)
Fly parasites	Filth flies	Animals	Brunetti (1981)
Praon spp.	Aphids	Potato	Shands et al. (1975)
Stethorus picipes Casey	Avocado brown mite	Avocado	McMurtry et al. (1969)
Trichogramma carverae (Oatman & Pinto)	Leafroller	Wine grapes	Glenn and Hoffman (1997)
Trichogramma spp.	Leafroller	Apple	Lawson et al. (1997)
Trichogramma pretiosum Riley	Bollworm, Tobacco budworm	Cotton	Stinner et al. (1974), Ables et al. (1982), King et al. (1985b)
Trichogramma minutum Riley	Tomato fruitworm, Cabbage looper, Tomato hornworm	Tomato	Oatman and Platner (1978)
	Imported cabbageworm, Cabbage looper	Cabbage	Parker (1971)
Trichogramma cacoecia Marchal	Sugarcane borer	Sugarcane	Jaynes and Bynum (1941)
	Codling moth	Apple	Dolphin et al. (1972)
Trichogramma nubiale Ertle & Davis	European corn borer	Corn	Kanour and Burbutis (1984)
Typhlodromus spp.	Cyclamen mite	Strawberry	Huffaker and Kennett (1956)
	McDaniel spider mite	Apple	Croft and McMurtry (1972)

reduction in larval infestations of the European corn borer following *Trichogramma* releases during the late 1970s in Germany (see also Bigler, 1983). Voronin and Grinberg (1981) reported reduction of pest species of *Loxostege* spp., *Agrotis* spp., and *Ostrinia* spp. following *Trichogramma* releases. In China a significant reduction in populations was reported for *Ostrinia* spp., *Heliothis* spp., and *Cnaphalocrocis* spp. and crop damage was also reduced (Li, 1984).

Oatman and Platner (1985) found that two common lepidopterous pests of avocado in southern California, *Amorbia cuneana* Walsingham and the omnivorous looper, *Sabulodes aegrotata* (Guenée), could be effectively controlled by releases of 50 000 *Trichogramma platneri* Nagarkatti in each of four uniformly spaced trees per acre. At least three weekly releases were required for control of *S. aegrotata*, while two were necessary for *A. cuneana*. Reduction in insect numbers and crop damage in several agroecosystems was also reported from China and the former USSR (Voronin and Grinberg, 1981; Li, 1984). Oatman and Platner (1971, 1978) demonstrated the technical feasibility of augmenting *Trichogramma pretiosum* Riley populations to reduce damage in tomato caused by the tomato fruitworm, *Helicoverpa zea* (Boddie); cabbage looper, *Trichoplusia ni* (Hübner), and *Manduca* spp., although these authors found that chemical control was necessary for pests not attacked by *T. pretiosum*. Oatman *et al.* (1983) reported studies conducted to develop an integrated control program for the tomato fruitworm and other lepidopterous pests on summer plantings of fresh-market tomatoes in southern California in 1978 and 1979. A regimen of twice-weekly applications of Dipel® (δ-endotoxin of *Bacillus thuringiensis* Berliner var. *kurstaki*), plus twice-weekly releases of *T. pretiosum*, was compared with weekly applications of methomyl. There were no significant differences in fruit yield or size between the two treatment programs. Methomyl adversely affected predator populations, host eggs, and egg parasitization by *T. pretiosum*, whereas Dipel® did not.

Inundative releases of *Trichogramma* spp. within commercial vineyards were used in Australia to assess effectiveness in controlling the tortricid leafroller, *Epiphyas postvittana* (Walker) in wine grapes (Glenn and Hoffman, 1997). Approximately 75 percent of the egg masses in the vineyard could be parasitized by *T. carverae* at release rates of 70 000 *Trichogramma* per hectare. This led to isolation of a *T. carverae* strain for putative commercial use. Inundative releases of *Trichogramma platneri* Nagarkatti resulted in significantly greater parasitism of obliquebanded leafroller, *Choristoneura rosaceana* (Harris), egg masses than in control plots (Lawson *et al.*, 1997).

Trichogramma spp. have been utilized in the Netherlands for biological control of lepidopteran pests. Two approaches were taken, the selection of "best" species and strains of *Trichogramma* (Van Lenteren *et al.*, 1982) and studies of the manipulation of *Trichogramma* behavior. The first approach has been studied extensively, especially in *Brassica* spp. (Pak *et al.*, 1988). Inundative releases of *Trichogramma* were feasible for control of *Mamestra brassicae* (Linnaeus) on Brussels sprouts, but control was not very effective at low host densities (van der Schaff *et al.*, 1984). Glas *et al.* (1981) also reported reduction in larval infestation of *Plutella xylostella* (Linnaeus) in cabbage crops. Van Heiningen *et al.* (1985) summarized several years' results of experimental releases of *Trichogramma*. The second approach to *Trichogramma* manipulation involved examination of semiochemical-mediated behavior. These studies indicated that kairomones and volatile substances released by adult female hosts were important in foraging behavior of *Trichogramma* spp. (Noldus and van Lenteren, 1983; Noldus *et al.*, 1988a,b). Preintroductory evaluation as outlined by Wackers *et al.* (1987) may improve prospects for success of augmentative release of strains of *Trichogramma* spp.

Although control can be achieved by augmentation with *Trichogramma* spp. under certain circumstances, variable results and insufficient pest control have also been reported (King *et al.*, 1985b). *Trichogramma pretiosum* Riley was tested in augmentative releases in Arkansas in 1981–82 and in North Carolina in 1983 for management of *H. zea* and *Heliothis virescens* (Fabricius) in cotton. These releases failed to provide adequate control in 1981–82. However, in 1983, cotton fields treated with seven augmentative releases of *T. pretiosum* at 306 000 emerged adults/ha/release yielded significantly more cotton than check fields (not insecticide treated). Insecticidal control fields yielded more cotton than check fields or *T. pretiosum* release fields, leading these researchers to conclude that management of *Heliothis* spp. in cotton by augmentative releases of *T. pretiosum* was not economically feasible at the time (King *et al.*, 1985b).

Consistent results in the use of *Trichogramma* necessitate the release of large numbers under controlled conditions. In addition to numbers released, the effectiveness of *Trichogramma* may also be influenced by

- Population density or dynamics of the pest
- Species or strain of *Trichogramma* released
- Vigor of the parasitoid released
- Method of distribution
- Crop phenology
- Number of other biological agents present
- Proximity to crops receiving insecticide and drift of insecticides into *Trichogramma* release areas (King *et al.*, 1985b).

Trichogramma spp. appear highly susceptible to most chemical insecticides, and lethal effects may result from direct exposure to spray application, drift, or posttreatment contact with pesticide residues on foliage (Bull and Coleman, 1985). Chemical insecticides may have contributed to inconsistency in results obtained in Arkansas and North Carolina from augmentation efforts (King *et al.*, 1985b).

King *et al.* (1985b) cited costs for release of *T. pretiosum* to control *Heliothis* spp. at US$7.68/ha per application. This cost compares well with the cost of a commonly used pyrethroid, fenvalerate, at US$16.18/ha when applied at 0.11 kg/ha. The pyrethroid needs to be applied only once for every two to three parasitoid applications. Nevertheless, the development of resistance to pyrethroids and other classes of insecticides by *Heliothis* (Luttrell *et al.*, 1987; Elzen *et al.*, 1990; Elzen, 1991; Elzen *et al.*, 1992) and to dimethoate by tarnished plant bug, *Lygus lineolaris* (Palisot de Beauvois) (Snodgrass and Scott, 1988), and further development of resistance, makes augmentation attractive.

Control of *Bemisia*

Widespread infestation by the silverleaf whitefly, *Bemisia argentifolii* Bellows & Perring, has caused severe economic damage in crops in the southwestern United States and northern Mexico. Heinz *et al.* (1994) found that releases of the coccinellid beetle, *Delphastus pusillus* (LeConte), in Imperial Valley cotton trials reduced immature whitefly densities to one-third of the densities found in cotton plots receiving no releases. It was suggested that this predator could be used widely to suppress silverleaf whitefly.

Parasitoids have also been evaluated for control of silverleaf whitefly. Biological control of silverleaf whitefly was undertaken using augmentative releases of commercially available *Encarsia formosa* Gahan (Parrella *et al.*, 1991). Although a dramatic decrease in the whitefly population was observed in the biological control area, it was not sufficient to prevent whitefly egg deposition on terminal growth of poinsettia, which is the harvestable product. Releases of *Encarsia luteola* Howard were evaluated for control of *B. argentifolii* in greenhouse-grown poinsettias. Whitefly densities in nonrelease cages were significantly greater than whitefly densities in release cages, indicating a significant impact on infestations. However, the cost of control with *E. luteola* was 5 times the cost of control with insecticides (Heinz and Parrella, 1994). Simmons and Minkenberg (1994) used a field cage evaluation of *Eretmocerus* nr. *californicus* in cotton infested with silverleaf whitefly. Seed cotton yield was significantly greater in high parasitoid release cages in comparison with those of low parasitoid release and control cages. They concluded that *E.* nr. *californicus* may be a useful control agent in an augmentative release strategy against whitefly on cotton. The effectiveness of inundative releases of *E. formosa* for control of silverleaf whitefly on poinsettia was determined in replicated experimental greenhouses (Hoddle *et al.*, 1997). A release rate of 3 wasps per plant per week provided better control of silverleaf whitefly than a low release and this was attributed to in-house wasp production. Hoddle and van Driesche (1999) found that *Eretmocerus eremicus* Rose & Zolnerowich effectively controlled silverleaf whitefly on poinsettia in greenhouses, but the use of *E. eremicus* was 44 times more expensive than using the insecticide imidacloprid.

The compatibility of abamectin with *E. formosa* as a part of integrated pest management of the greenhouse whitefly, *Trialeurodes vaporariorum* Westwood, was assessed in a series of laboratory and greenhouse experiments (Zchori-Fein *et al.*, 1994). The percentage of parasitism did not differ significantly among poinsettia plants treated with and without abamectin. In laboratory experiments, more than 50 percent of the *E. formosa* eclosed from parasitized whitefly pupae on bean leaves dipped in formulated abamectin. They suggested that abamectin might be used to reduce whitefly numbers on poinsettia without eliminating the parasitoid population when releases of *E. formosa* are not satisfactory.

Other species

Inundative releases of the parasitoid *Diglyphus begini* (Ashmead) reduced *Liriomyza trifolii* (Burgess) populations in greenhouse marigolds grown for seed to approximately

---

zero within 8 weeks of the first release and they remained at that level until harvest. Releases of *E. formosa* and *Chrysoperla carnea* (Stephens) may have been responsible for the maintenance of *T. vaporariorum*, *Aphis gossypii* Glover, and *Myzus persicae* (Sulzer) at acceptable levels in three of four trials (Heinz and Parrella, 1990).

Laboratory-reared adults of the tachinid parasitoid *Archytas marmoratus* (Townsend) were released and evaluated against larvae of *H. zea* in whorl corn. Data suggested that *A. marmoratus* females were highly efficient in finding larvae of *H. zea*, and that rates of parasitism observed in release areas were likely lower than could be expected from sustained releases throughout corn ecosystems. It was suggested that development of economical methods of mass propagation of *A. marmoratus* could provide safe, efficient, and selective management of the early-season populations of *H. zea* in whorl stage corn (Gross, 1990).

Laboratory-reared *Cotesia melanoscela* (Ratzeburg) was released sequentially over 3 weeks at an average level of 12 000 females per hectare in three isolated mixed-hardwood woodlots infested with gypsy moth. Significantly higher rates of generational parasitism were achieved in release woodlots (15.4 percent) than in control woodlots (5.1 percent). Despite significantly higher rates of parasitism in release woodlots, inundative releases of *C. melanoscela* failed to reduce gypsy moth populations as determined from egg mass counts (Kolodny-Hirsch *et al.*, 1988).

Females of the predacious mite, *Noeseiulus fallacis* (Garman), were released at point locations in strawberry for control of two-spotted spider mite, *Tetranychus urticae* Koch (Croft and Coop, 1998). Models for release rates and dispersal were developed from the data. Grafton-Cardwell and Ouyang (1995) made single releases of another predacious mite, *Euseius tularensis* Congdon, in the spring of 1992 in two citrus orchards in the San Joaquin Valley of California. A greater cumulative number of predacious mites was significantly correlated with improved biological control of citrus thrips, *Scirtothrips citri* (Moulton), as measured by reduced scarring of fruit in the orchards. *E. tularensis* could be important in reducing citrus thrips in citrus groves managed under a selective pesticide program.

The impact of repeated releases of the egg parasitoid *Anaphes iole* (Girault) on the development of *Lygus hesperus* Knight populations and fruit damage in commercial strawberry fields was evaluated (Norton and Welter, 1996). Following the release of 37 000 parasitoids per hectare, 50 percent of *L. hesperus* eggs were parasitized. These results demonstrated that mass releases of *A. iole* reduced the development of populations of *L. hesperus* nymphs and the subsequent fruit damage, and thus this parasitoid may prove to be an effective management tool for this pest.

Convergent lady beetles, *Hippodamia convergens* Guérin-Méneville, collected from aggregations in California were released for control of melon aphids, *Aphis gossypii* Glover, infesting potted chrysanthemum, *Dendranthema grandiflora* (Tzelev) "Hurricane," outdoors. The results showed that release of commercially available convergent lady beetles can provide augmentative control of relatively high aphid densities on small plotted plants (Dreistadt and Flint, 1996).

Daane *et al.* (1996) examined the effectiveness of inundative release of common green lacewing, *Chrysoperla carnea* (Stephens), to suppress two vineyard pests, *Erythroneura variabilis* Beamer and western grape leafhopper, *E. elegantula* Osborn, in small-plot and on-farm trials. The average reduction of leafhoppers in *C. carnea*-release plots, as compared with no-release plots, was 29.5 percent in cages, 15.5 percent in three-vine plots, and 9.6 percent in commercial vineyards. A significant, although only weakly positive, correlation was found between release rate and effectiveness. There was also a greater reduction of leafhopper nymphs when lacewings were released as larvae, as compared with eggs. Combining data from all studies, the number and percentage of reductions of leafhopper nymphs were related to leafhopper density. Most importantly, when leafhopper densities were above the suggested economic injury level (15–20 nymphs per leaf), the reduction in leafhopper number was frequently not sufficient to lower the leafhopper density below the economic injury threshold.

Catolaccus grandis (Burks)

In 1988, E.G. King hypothesized that one of our most intractable pests, the boll weevil, could be controlled by augmentative releases of selective parasitoids. He further hypothesized that failure to become established, as in the case of exotic parasitoids of the boll weevil, was not critical in an augmentative release program. In fact, he concluded that population densities of boll weevils tolerated by cotton growers, in season, are so low that they could not support naturally occurring parasitoid populations. These hypotheses were documented in two USDA, Agricultural Research Service CRIS Work Projects. These hypotheses were further outlined by Knipling (1992), who developed a theoretical model postulating the suppressive effects of a host-specific parasitoid released against the boll weevil.

The selective parasitoid *Catolaccus grandis* (Burks) (Hymenoptera: Pteromalidae) was chosen for release against the boll weevil in the 1989 ARS Work Planning Session. Previous attempts to import and establish *C. grandis* on the boll weevil were unsuccessful, but it was demonstrated that the parasitoid was preadapted to the in-season biotic and abiotic environment of the cotton agroecosystem (Johnson *et al.*, 1973; Cate *et al.*, 1990). *C. grandis* effectively searched for boll weevil-infested buds on the ground and on the plant.

The technical feasibility of suppressing the boll weevil by augmentative releases of insectary-produced *C. grandis* (Morales-Ramos *et al.*, 1992; Roberson and Harsh, 1993) was demonstrated in 1992 (Summy *et al.*, 1992, 1995) and 1993 (Morales-Ramos *et al.*, 1994; Summy *et al.*, 1994). Morales-Ramos *et al.* (1993) reported a mathematical model simulating the effects of releases of *C. grandis* on the boll weevil. This model theoretically demonstrated that high rates of parasitism could eliminate in-field reproduction by the boll weevil and preclude the need for subsequent insecticide treatments. King *et al.* (1993) outlined a strategy for integrating augmentation into short-season cotton production. Summy *et al.* (1994) experimentally demonstrated this integration in 1993, but chemical insecticides were still not applied for early-season insect pests, viz., invading overwintered boll weevils and cotton fleahoppers. Consequently, there remained a numerical, though not statistical, difference in lint yield between the IPM-treated control fields and the parasitoid-release fields. Additionally, the experimental cotton fields were not managed as commercial cotton fields. Accordingly, additional experiments were conducted to demonstrate that the boll weevil could be suppressed to subeconomic levels in commercially grown cotton by the aid of augmentative releases of *C. grandis* during the boll weevil F_1 and F_2 generations as immatures in buds on the plant and ground.

Subsequently, King *et al.* (1995) reported the feasibility of large-scale rearing of *C. grandis* on artificial-diet-reared boll weevil third instars and augmentative releases in commercially managed cotton fields in South Texas cotton fields to suppress boll weevils. Up to 90 percent mortality of boll weevil third instars and pupae was recorded in three parasitoid-release fields, but mortality was substantially less in three IPM-treated control fields. Survival of boll weevil third instars and pupae during the F_2 and F_3 generations was near to undetectable in two of the parasitoid-release fields and was reduced in the third parasitoid-release field. Lint yield was significantly higher in the parasitoid-release fields (mean = 1047 lb lint/acre; 1173.5 kg/ha) than in the IPM-treated control (mean = 929 lb lint/acre; 1041.25 kg/ha), and the number of chemical insecticide applications was reduced from 11 to 5. Expansion of *C. grandis* from field-by-field early-season augmentative releases to early-season area-wide releases is projected to be substantially more powerful as a population suppressant, preempting the development of damaging boll weevil populations.

Sensitivity analysis was done to estimate costs of using *C. grandis* over a range in values for selected factors. A base case assumed values for the following factors: three 100 000 acre (~40 500 ha) tracts of cotton treated sequentially, a US$2.1 million facility amoritized over 10 years at 9 percent interest with an artificial diet (Rojas *et al.*, 1999) at cost of US$2.35 per 1000 wasps. Economic analyses projected an estimated total seasonal cost of US$23 per acre for producing and releasing *C. grandis* for boll weevil suppression, with a two-thirds reduction in insecticide runoff and a one-half reduction in insecticide leaching (Ellis *et al.*, 1997).

Evaluating natural enemies for augmentation

Effective manipulation of natural enemies requires a thorough knowledge of their biology and host associations. Background information may be gained through laboratory studies. However, it is imperative that such data be followed by extensive biological studies in the field. A comparison of laboratory, field cage, and field studies would provide useful information that could be used to predict the impact of the natural enemy on pest populations. Ecological studies may help to determine the influence of habitat, biotypes, or semiochemicals on parasitoid foraging, and may provide helpful information for determining in which habitat to release natural enemies.

The level of control achieved in an augmentation program may be a function of a myriad of interacting factors (Huffaker *et al.*, 1977). Primary factors include the availability of hosts and host–parasitoid synchrony, conditions of weather during release of natural enemies, including the effects of environmental factors on foraging,

Gary W. Elzen, Randy J. Coleman, and Edgar G. King

influence of habitat type, chemical pesticide usage, and fitness of laboratory-reared parasitoids. Thus, experiments must be conducted that will determine specific situations in which biological control by augmentation may work. All parameters, biotic and abiotic, should be explored in evaluating augmentation results. Augmentation of natural enemies of row crop pests may be implemented only after considerable effort has been expended to prove the feasibility of this approach. Thereafter, the efficiency and economic benefits must be determined. Reliance upon the use of entomophagous arthropods to control pests should be limited to those situations where scientifically, environmentally, and economically sound procedures are available. In addition, fundamental knowledge of search rates, functional response, and efficiency could significantly add to the predictability of success in augmentation efforts.

Sabelis and Dicke (1985) and van Lenteren (1986) have advocated screening of the biological characteristics of natural enemies for biological control. Parasitism and predation can vary spatially owing to variation in parasitoid host plant detection, search rate, or retention of parasitoids on host plants. Host plant species and stage, host density, and weather are likely to affect all three processes. Parasitism can vary temporally because of variation in host detection, search, retention, and parasitoid natality and mortality. Research designed to gather information to predict distribution of parasitism across host plants under varying conditions could yield important information on the population dynamics of hosts and parasitoids. These predictions are crucial to rational conservation and augmentation of parasitoids. Variation in parasitism found in host plant surveys may arise from variation in attraction or retention of wasps by semiochemicals directly or indirectly derived from the host plants. Studies of host habitat preference may provide clues to the optimum habitat and conditions in which to release parasitoids in augmentation programs.

Field experiments may provide insights into the efficiency of a particular parasitoid. A thorough understanding of parasitoid–host interactions may aid augmentation efforts.

Theoretically, the means to increase the suppressive effect of a natural enemy can be found by increasing their numbers through propagation and release. The rate of establishment may be increased by releasing larger numbers (Beirne, 1975; Ehler and Hall, 1982) but this is not always the case (Greathead, 1971). Field evaluations must provide the data necessary for defining the number of predators or parasitoids required for release per unit area and this, along with mass production and distribution technology, will determine the economic feasibility of the approach.

Introduced natural enemies, with actual or projected use to augment natural populations, should be evaluated by the criteria proposed by Luck *et al.* (1988). Experimental evaluation through life table analysis, examination of percentage parasitism, key factor analysis, and the use of simulation models, may provide insights into the probability of success. Evaluation of natural enemies may include introduction and augmentation, and techniques using cages and barriers, removal of natural enemies, prey enrichment, direct observation, and biochemical evidence of natural enemy feeding, in quantifiable experiments to gauge the impact of the natural enemies (Luck *et al.*, 1988).

Conservation

An important issue affecting the successful augmentation of natural enemies is the use of pesticides. The effects of pesticides on beneficial insects have been considered in detail (Johnson and Tabashnik, 1999; Tabashnik and Johnson, 1999) and have also been reviewed by Croft and Brown (1975), Croft and Morse (1979), Croft (1990), and Jepson (1989). Recommendations for changing control practices to preserve natural enemies were addressed by Plapp and Bull (1989). King and Coleman (1989) reviewed insecticide use in cotton and the value of predators and parasites for managing *Heliothis*. Elzen (1989) reviewed the sublethal effects of pesticides on parasitoids. Thus, numerous studies have documented the detrimental effects of pesticides on natural enemies and it is important to note that an underlying problem in the practical implementation of augmentation is the use of pesticides.

Monitoring

Monitoring and sampling of natural enemies that are released in augmentation programs may provide useful information for models to predict the impact of entomophages on reducing herbivore-induced damage or plant stress. Modeling of population interactions requires accurate tools to determine absolute densities of beneficials and pests. Monitoring of beneficial insect populations, particularly parasitoids, may be complicated by factors such as lack of a stable sex ratio, movement (because females must forage for often patchily distributed hosts), weather factors, and lack of synchrony with host populations.

Insect control guidelines often recognize the impact of natural enemy populations on pest populations (Rude, 1984). However, explicit instructions for using natural enemies in decision making are generally lacking, and where present are used with reservation. However there are exceptions:

1. Michelbacher and Smith (1943) recommended that insecticide control decisions in alfalfa for *Colias eurytheme* Boisduval be made only after determining that the number of *Apanteles medicaginis* Muesebeck present was incapable of maintaining the pest under control.
2. Croft and McMurtry (1972) reported on a decision-making index for predicting the probability of adequate control of a phytophagous mite that would occur depending on the predator–prey ratio per apple leaf.
3. Drees (1984) described a shake-bucket method for sampling predators in cotton combined with terminal inspection for *Heliothis* eggs. No action (no insecticide treatment) was recommended when the predator-to-*Heliothis* ratio was 1:1 or greater.
4. Young and Willson (1984) developed a model to predict damage reduction to flower buds or fruit by *Heliothis* spp. in the absence or presence of two coleopteran predators. They reported that in Oklahoma cotton if one or more coleopteran predators (convergent lady beetle or soft flower beetle) are present per 0.8 row meter, then *Heliothis* damage is unlikely to occur.

Gary W. Elzen, Randy J. Coleman, and Edgar G. King

Rearing

Efficient and cost-effective methods of rearing entomophagous arthropods must be developed if augmentative releases are to be feasible (Beirne, 1974). The importance of adequate and cost-efficient culture methods is particularly evident in greenhouse programs (van Lenteren, 1986). The nutrition and culture of entomophagous insects was recently reviewed by Thompson (1999a). In addition, large numbers of beneficial insects may be employed in glasshouse, field cage, and laboratory studies. Literally thousands of beneficial organisms, available at somewhat unpredictable times, may be required for commercial augmentation releases. Finney and Fisher (1964) discuss problems associated with the culture of entomophagous insects. Considerable attention has been devoted to development of techniques to produce quality insects in large numbers (Smith, 1966; King and Leppla, 1984; Anderson and Leppla, 1991). The genetic implications of long-term laboratory rearing of insects are addressed elsewhere (Mackauer, 1972, 1976; Bouletreau, 1986). Powell and Hartley (1987) described techniques for producing large numbers of parasitoids efficiently. These authors adapted a multicellular host-rearing tray technique (Hartley *et al.*, 1982) to rear *Microplitis croceipes* (Cresson) and several other species of parasitoids. Techniques reduced parasitoid harvest time by one-half, and simultaneous release of nearly 17 000 wasps was possible using low temperature storage. Morales-Ramos *et al.* (1992) reported methods for mass rearing of *C. grandis* and recently reported a new system to more efficiently mass-rear *C. grandis* using a novel cage system (Morales-Ramos *et al.*, 1997). Powell and Hartley (1987) noted several factors that were important for maintaining a large-scale rearing program and that may be applicable to other programs:

- Continuous host supply
- Use of environmental chambers to alter developmental rates of hosts and parasitoids
- Constant and appropriate environmental conditions
- Sanitary rearing conditions with flash sterilization of diet
- Use of laminar flow hoods
- Autoclaving reusable supplies
- Disinfection of work areas
- Acid or antibiotics in water or food
- Adequate technical support, space, and equipment.

The quality of mass-reared insects should also be assessed (Boller and Chambers, 1977). Beneficial insects that have been reared for augmentative releases are shown in Table 3.2.

In Vitro Rearing

Commercialization of augmentation may be practical only for selected organisms for which suitable diets and storage methods are developed. Artificial rearing may increase practical use of augmentation for widespread pest control. Fifty-eight species of natural enemy have been reared *in vitro*. Forty-three species of Hymenoptera and

Table 3.2 Entomophagous arthropods that have been reared for augmentation (noncommercial sources)

Entomophage	Target pest	Commodity	Production host	References
Parasites				
Aphytis lingnanensis Compere	Red Scale	Citrus	Oleander scale	King and Morrison (1984)
Other red scale parasites	Red scale	Citrus	Oleander scale	Brunetti (1981)
Aphidius spp.	Pea aphid	Peas	Pea aphid	Ridgway et al. (1977)
Archytas marmoratus Townsend	*Helicoverpa zea* (Boddie)	Corn	*H. zea*	Gross (1990)
Bracon kirkpatricki (Walker)	Pink bollworm	Cotton	Beet armyworm	Morrison and King (1977)
Catolaccus grandis (Burks)	Boll Weevil	Cotton	Boll weevil	King et al. (1995)
Cotesia melanoscela (Ratzburg)	Gypsy moth	Hardwood	Gypsy moth	Kolodny-Hirsch et al. (1988)
Microchelonus blackburni Cameron	Pink bollworm	Cotton	Not given	Brunetti (1981)
Diglyphus begini (Ashmead)	*Liriomyza trifolii* (Burgess)	Marigolds	*L. trifolii*	Heinz and Parrella (1990)
Encarsia formosa Gahan	Greenhouse whitefly	Not given	Not given (probably greenhouse whitefly)	Brunetti (1981)
				King and Morrison (1984)
E. formosa	Greenhouse whitefly	Poinsettia	Greenhouse whitefly	Zchori-Fein et al. (1994)
E. luteola Howard	*Bemisia argentifolii*	Poinsettia	*B. argentifolii*	Heinz and Parrella (1994)
	Bellows & Perring			
Eretmocerus eremicus Rose &	*B. argentifolii*	Poinsettia	*B. argentifolii*	Hoddle and van Driesche (1999)
Zolnerowich				
Trichogramma dendrolimi	*Dendrolimus punctata*	Not given	*In vitro*	Dai (1988)
Trichogramma confusum	*Chilo sacchariphagus*	Not given	*In vitro*	Dai (1988)
	Chilo infuscatellus			
T. carverae (Oatman & Pinto)	*Epiphyas postvittana* (Walker)	Wine grapes	*Sitotroga cerelalella* (Oliver)	Glenn and Hoffman (1997)
Fly parasites	Filth flies	Man, other animals	House fly	Brunetti (1981)
Lixophaga diatraeae Townsend	Sugarcane borer	Sugarcane	Greater wax moth	King et al. (1979)
Lysiphlebus testaceipes (Cresson)	Greenbug	Sorghum	Greenbug	Morrison and King (1977)
Macrocentrus ancylivorus Rohwer	Oriental fruit moth	Fruit	Potato tuberworm	Morrison and King (1977)
Metaphycus helvolus (Compere)	Black scale	Citrus	Black scale	Morrison and King (1977)
Pediobius foveolatus (Crawford)	Mexican bean beetle	Soybean	Mexican bean beetle	Ridgway et al. (1977)
Trichogramma spp.	Lepidopterous pests	Several crops	Angoumois grain moth	Morrison and King (1977)
Predators				
Chrysopa carnea Stephens	Several	Several crops	Angoumois grain moth	Morrison and King (1977)
Cryptolaemus montrouzieri Mulsant	Citrus mealybug	Citrus	Citrus mealybug	Morrison and King (1977)
Predatory mites	Spider mites	Several crops	Two-spotted spider mites	Morrison and King (1977)
				Brunetti (1981)
Delphastus pusillus Gahan	*B. tabaci*	Cotton	*B. tabaci*	Heinz et al. (1994)
Euseius tularensis Congdon	*Scirtothrips citri* (Moulton)	Citrus	*Phaseolus limensis*	Grafton-Cardwell and Ouyang (1995)
Neoseiulus fallacis (Garman)	Spider mites	Strawberry	*P. lunatus*	Croft and Coop (1998)

15 species of Diptera have been cultured with varying success (Thompson, 1999b). Predators, notably *Chrysoperla carnea* (Stephens), have been reared on artificial diets (Vanderzant, 1973; Martin *et al.*, 1978). While there have been numerous successes in oviposition stimulant identification or partial rearing (Nettles and Burks, 1975; Nettles, 1982), definitive development of a feasible *in vitro* rearing system for beneficials has yet to be developed. Presently most parasitoids are expensive to rear, and the costs involved would preclude mass rearing. Although considerable advances have been made in *in vivo* rearing, the advances have not been achieved with *in vitro* rearing to such an extent. The work of Wu *et al.* (1982) illustrated an instance in which a completely synthetic artificial host egg was produced that contains no insect derivatives and supports *Trichogramma* oviposition and development. Greany *et al.* (1984) suggested that mass rearing of *Trichogramma* using completely artificial hosts would become economical within the year 1989, but this prediction was not realized.

C. *grandis* has been successfully reared on artificial diet (Guerra *et al.*, 1993; Guerra and Martinez, 1994). Methods were recently developed to consistently produce *C. grandis* of higher quality (Rojas *et al.*, 1996) and at a lower cost (Rojas *et al.*, 1997). An improved meridic diet for rearing *C. grandis* was recently patented (Rojas *et al.*, 1999). *C. grandis* reared by the methods of Rojas *et al.* (1996) were recently evaluated in laboratory and field studies in Texas (Morales-Ramos *et al.*, 1998). Females reared on boll weevils had a higher pupal weight and fecundity than females reared *in vitro*, but females reared *in vitro* exhibited significantly higher survival during the period of most intensive reproductive activity. The movement and searching capacity under field conditions of the *in vitro* and *in vivo* reared *C. grandis* were compared. Dispersal ability and searching capacity were not significantly different within a 30 m radius from a release point. Nevertheless, no significant difference in boll weevil mortality induced by parasitism was recorded between the two methods. The use of artificial diets is a promising method for mass-propagating *C. grandis* (Ellis *et al.*, 1997).

Hymenopterous larval endoparasitoids have not been successfully reared to the adult stage on artificial diet. However, *Cotesia marginiventris* (Cresson) and *M. croceipes* have been reared on artificial media through the first instar (Greany, 1986). Hu and Vinson (1998) reported development of *Campoletis sonorensis* (Cameron) to mature fifth instar larvae. Larval endoparasitoids have evolved complex mechanisms that interact with the host's internal dynamics and organs without damaging this environment or causing untimely demise of the host. The workings of these interacting factors must be understood for *in vitro* rearing of larval endoparasitoids to become a reality. Developments in artificial rearing of beneficial insects on artificial diet may allow production of sufficient numbers of beneficials to feasibly implement the further evaluation of natural enemies as biological control agents.

Parasitoids that have been reared to adults on artificial media include the larval ectoparasitoid *Exeristes roborator* (Fabricius) (Thompson, 1982) and endoparasitoids of eggs of *T. pretiosum* (Hoffman *et al.*, 1975) and *T. dendrolimi* (Wu *et al.*, 1982), larvae of *Lixophaga diatreae* (Towns) (Grenier *et al.*, 1978) and *Eucelatoria bryani* Sabrosky (Nettles *et al.*, 1980), and pupae of *Brachymeria lasus* (Walker) (Thompson, 1983), *Pachycrepoideus vinclemia* Rondani (Thompson *et al.*, 1983), *Itoplectis conquisitor* (Say) (House, 1978), *Pteromalus puparum* Linnaeus (Hoffman and Ignoffo, 1974), and *C. grandis* (Rojas *et al.*, 1996).

Conclusions

Recommendations for augmentation biological control studies:

- Modify chemical control practices to preserve natural enemies.
- Develop selective pesticides favoring arthropod natural enemies.
- Integrate biological control with cultural practices, habitat manipulation, and host plant resistance.
- Develop methodology for estimating abundance and impact of natural enemies.
- Integrate the natural enemy component into crop/pest models; conduct studies to determine factors that influence fitness (i.e., reproductive success) of released natural enemies.
- Identify and colonize the best biotypes of predators and parasites including those of foreign origin.
- Increase efforts on genetic improvement of natural enemies.
- Develop *in vitro* rearing techniques for natural enemies.

After these criteria have been studied, it may become apparent that a particular augmentation effort has the chance for success. At that time an entire, rational pest management package would need to be developed and a pilot program could be implemented. Ultimately, the utilization of natural enemies in pest control will require a series of basic and applied research steps.

References

Ables, J.R. and Ridgway, R.L. (1981) Augmentation of entomophagous arthropods to control pest insects and mites. In G.C. Papavisas (ed.), *Biological Control in Crop Protection*, Osman and Co., Allenheld, Osman & Co., Totowa, NJ, pp. 273–303.

Anderson, T.E. and Leppla, N.C. (1991) *Advances in Insect Rearing for Research and Pest Management*, Westview Studies in Insect Biology, Westview Press/Oxford and IAH Publishers, Boulder, CO.

Beirne, B.P. (1974) Status of biological control procedures that involve parasites and predators. In F.G. Maxwell and F.A. Harris (eds.), *Proceedings of the Summer Institute on Biological Control of Plant Insects and Diseases*, University Press of Mississippi, Jackson, MS, pp. 69–76.

Beirne, B.P. (1975) Biological control attempts by introductions against pest insects in the field in Canada. *Can. Entomol.*, **107**, 225–236.

Beglyarov, G.A. and Smetnik, A.I. (1977) Seasonal colonization of entomophages in the U.S.S.R. In R.L. Ridgway and S.B. Vinson (eds.), *Biological Control by Augmentation of Natural Enemies*, Plenum Press, New York, pp. 283–328.

Bigler, F. (1983) Erfahrungen bei der biologischen Bekanpfung des Maiszunslers mit *Trichogramma*—Schlupfwespen in des Scheveiz. *Mitt. Schweiz. Landwirtsch.*, **31**, 14–22.

Boller, E.F. and Chambers, D.L. (1977) Quality aspects of mass-reared insects. In R.L. Ridgway and S.B. Vinson (eds.), *Biological Control by Augmentation of Natural Enemies*, Plenum Press, New York, pp. 219–235.

Bouletreau, M. (1986) The genetic and coevolutionary interactions between parasitoids and their hosts. In J. Waage and D. Greathead (eds.), *Insect Parasitoids*, Academic Press, New York, pp. 169–200.

Brunetti, K.M. (1981) *Suppliers of Beneficial Organisms in North America*, California Dept. of Food and Agriculture, Biological Control Services Program, Sacramento, CA.

Brunson, M.H. and Allen, A.W. (1944) Mass liberation of parasites for immediate reduction of oriental fruit moth injury to ripe peaches. *J. Econ. Entomol.*, **37**, 411–416.

Bryan, D.E., Fye, R.E., Jackson, C., Cate, J.R., Krauter, P.C. and Godfrey, K.E. (1973) Pests of cotton. In D.H. Habect, F.D. Bennett and J.H. Frank (eds.), *Classical Biological Control in the Southern United States*, South. Coop. Ser. Bull. No. 335.

Bull, D.L. and Coleman, R.J. (1985) Effect of pesticides on *Trichogramma* spp. *Supp. Southwest Entomol.*, **8**, 156–168.

Cate, J.R., Krauter, P.C. and Godfrey, K.E. (1990) Pests of cotton. In D.H. Habeck, F.D. Bennett and J.H. Frank (eds.), *Classical Biological Control in the Southern United States*, South Coop. Ser. Bull No. 335, 17–29.

Chant, D.A. (1959) Phytoseiid mites (Acarina: Phytoseiidae). Part 1. Bionomics of seven species in south-eastern England. *Can. Entomol.*, **12**, 1–44 (suppl).

Cooke, W.C. (1963) Ecology of the pea aphid in the Blue Mountain area of eastern Washington and Oregon. *USDA Tech. Bull. 1287*, 48pp.

Croft, B.A. (1990) *Arthropod Biological Control Agents and Pesticides*, Wiley, New York.

Croft, B.A. and Brown, A.W.A. (1975) Responses of arthropod natural enemies to insecticides. *Annu. Rev. Entomol.*, **20**, 285–325.

Croft, B.A. and Coop, L.B. (1998) Heat units, release rate, prey density, and plant age effects on dispersal by *Neoseiulus fallacis* (Acari: Phytoseiidae) after inoculation into strawberry. *J. Econ. Entomol.*, **91**, 94–100.

Croft, B.A. and McMurtry, J.A. (1972) Minimum releases of *Typhlodromus occidentalis* to control *Tetranychus mcdanieli* on apple. *J. Econ. Entomol.*, **65**, 188–191.

Croft, B.A. and Morse, J.G. (1979) Research advances on pesticide resistance in natural enemies. *Entomophaga*, **24**, 3–11.

Daane, K.M., Yokota, G.Y., Zheng, Y. and Hagen, K.S. (1996) Inundative release of common green lacewings (Neuroptera: Chrysopidae) to suppress *Erythroneura variabilis* and *E. elegantula* (Homoptera: Cicadellidae) in vineyards. *Environ. Entomol.*, **25**, 1224–1234.

Dai, K.J. (1988) Research and utilization of artificial host egg for propagation of parasitoid *Trichogramma*. *Colloques de l'INRA*, **43**, 311–318.

DeBach, P. (1964) Successes, trends, and future possibilities. In P. DeBach (ed.), *Biological Control of Insect Pests and Weeds,* Chapman and Hall, London, pp. 673–713.

DeBach, P. and Hagen, K.S. (1964) Manipulation of entomophagous species. In P. DeBach (ed.), *Biological Control of Insect Pests and Weeds*, Chapman and Hall, London, pp. 429–455.

Dolphin, R.E., Clevelenad, N.L., Mouzin, L.E. and Morrison, R.K. (1972) Releases of *Trichogramma minutum* and *T. cacoeciae* in an apple orchard and the effects on populations of codling moths. *Environ. Entomol.*, **1**, 481–484.

Drees, B.M. (1984) Management of cotton insects in south and east Texas counties. *Texas Agric. Ext. Serv.*, B-1204.

Dreistadt, S.H. and Flint, M.L. (1996) Melon aphid (Homoptera: Aphididae) control by inundative convergent lady beetle (Coleoptera: Coccinellidae) release on chrysanthemum. *Environ. Entomol.*, **25**, 668–697.

Ehler, L.E. and Hall, R.W. (1982) Evidence for competitive exclusions of introduced natural enemies in biological control. *Environ. Entomol.*, **11**, 1–4.

Ellis, J., Johnson, J., Chowdury, M., Edwards, R. and Lacewell, R.D. (1997) Estimated economic feasibiliity of *Catolaccus grandis* in control of the boll weevil. *Environmental Issues/Sustainability Team Technical Report 97–1*, Texas A&M Agriculture Program, College Station, TX.

Elzen, G.W. (1989) Sublethal effects of pesticides on beneficial parasitoids. In P.C. Jepson (ed.), *Pesticides and Non-Target Invertebrates*, Intercept, Wimborne, Dorset, UK, pp. 129–150.

Elzen, G.W. (1991) Pyrethroid resistance and carbamate tolerance in a field population of tobacco budworm in the Mississippi Delta. *Southwest Entomol. Suppl.*, **15**, 27–31.

Elzen, G.W., O'Brien, P.J. and Snodgrass, G.L. (1990) Toxicity of various classes of insecticides to pyrethroid-resistant *Heliothis virescens* larvae. *Southwest Entomol.*, **15**, 33–38.

Elzen, G.W., Leonard, B.R., Graves, J.B., Burris, E. and Micinski, S. (1992) Resistance to pyrethroid, carbamate, and organophosphate insecticides in field populations of tobacco budworm (Lepidoptera: Noctuidae) in 1990. *J. Econ. Entomol.*, **85**, 2064–2072.

Finney, G.L. and Fisher, T.W. (1964) Culture of entomophagous insects and their hosts. In P. DeBach (ed.), *Biological Control of Insect Pests and Weeds*, Chapman and Hall, London, pp. 328–355.

Fisher, T.W. (1963) Mass culture of *Cryptolaemus* and *Leptomastix*—natural enemies of citrus mealybug. *Calif. Agric. Exp. Stn. Bull. 797*.

Flanders, S.E. (1929) The mass production of *Trichogramma minutum* Riley and observation on the natural and artificial parasitism of the codling moth egg. *Transcripts, 4th Int. Congr. Entomol.*, 2, 110–130.

Flanders, S.E. (1930) Mass production of egg parasites of the genus *Trichogramma*. *Hilgardia*, 4, 464–501.

Fye, R.E. (1972) Manipulation of *Polistes exclamans arizonensis*. *Environ. Entomol.*, 1, 55–57.

Glas, P.C., Smits, P.H., Vlaming, V. and van Lenteren, J.C. (1981) Biological control of lepidopteran pests in cabbage crops by means of inundative releases of *Trichogramma* species: a combination of field and laboratory experiments. *Med. Fac. Landbonuw Rijkuniv. Gent.*, 46, 487–497.

Glenn, D.C. and Hoffmann, A.A. (1997) Developing a commercially viable system for biological control of light brown apple moth (Lepidoptera: Tortricidae) in grapes using endemic *Trichogramma* (Hymenoptera: Trichogrammatidae). *J. Econ. Entomol.*, 90, 370–382.

Grafton-Cardwell, E.E. and Ouyang, Y. (1995) Augmentation of *Euseius tularensis* (Acari: Phytoseiidae) in citrus. *Environ. Entomol.*, 24, 738–747.

Greany, P. (1986) *In vitro* culture of hymenopterous larval endoparasitoids. *J. Insect Physiol.*, 32, 409–419.

Greany, P.D., Vinson, S.B. and Lewis, W.J. (1984) Finding new opportunities for biological control. *Bioscience*, 34, 690–696.

Greathead, D.J. (1971) A review of biological control in the Ethiopian region. *CIBC Tech. Comm. 5*.

Greathead, D.J. (1986) Parasitoids in classical biological control. In J. Waage and D.J. Greathead (eds.), *Insect Parasitoids*, Academic Press, New York, pp. 289–318.

Grenier, S., Bonnet, G., Delobel, B. and Laviolette, P. (1978) Development en milieu artificiel du parasitoide *Lixophaga diatraeae* (Towns) (Diptera: Tachinidae). Obtention de l'imago a partir de l'oeuf. *C.R. Acad. Sci. Paris*, 287, 535–538.

Gross, H.R. (1990) Field release and evaluation of *Archytas marmoratus* (Diptera: Tachinidae) against larvae of *Heliothis zea* (Lepidoptera: Noctuidae) in whorl stage corn. *Environ. Entomol.*, 19, 1122–1129.

Guerra, A.A. and Martinez, S. (1994) An *in vitro* rearing system for the propagation of the ectoparasitoid *Catolaccus grandis*. *Entomol. Exp. Appl.*, 72, 11–16.

Guerra, A.A., Robacker, K.M. and Martinez, S. (1993) *In vitro* rearing of *Bracon mellitor* and *Catolaccus grandis* with artificial diets devoid of insect components. *Entomol. Exp. Appl.*, 68, 303–307.

Hagen, K.W., van den Bosch, R. and Dahlsten, D.L. (1971) The importance of naturally-occurring biological control in the western United States. In C.B. Huffaker (ed.), *Biological Control*, Plenum Press, New York, pp. 253–293.

Halhill, J.E. and Featherston, P.E. (1973) Inundative releases of *Aphidius smithi* against *Acyrthosiphon pisum*. *Environ. Entomol.*, 2, 469–472.

Hart, R.E. (1972) Compensatory releases of *Microterys flavus* as a biological control agent against brown soft scale. *Environ. Entomol.*, 1, 414–419.

Hartley, G.G., King, E.G., Brewer, F.D. and Gantt, C.W. (1982) Rearing of *Heliothis* sterile hybrid with a multicellular larval rearing container and pupal harvesting. *J. Econ. Entomol.*, 75, 7–10.

Hassan, S.A. (1982) Mass production and utilization of *Trichogramma*. 3. Results of some research projects related to the practical use in the Federal Republic of Germany. Les *Trichogramma*. *Colloques INRA*, 9, 213–218.

Heiningen, T.G. van, Pak, G.A., Hassan, S.A. and van Lenteren, J.C. (1985) Four years' results of experimental releases of *Trichogramma* egg parasites against lepidopteran pests in cabbage. *Med. Fac. Landbonuw. Rigksuniv. Gent.*, 50, 379–388.

Heinz, K.M. and Parrella, M.P. (1990) Biological control of insect pests on greenhouse marigolds. *Environ. Entomol.*, 19, 825–835.

Heinz, K.M. and Parrella, M.P. (1994) Biological control of *Bemisia argentifolii* (Homoptera: Aleyrodidae) infesting *Euphorbia pulcherrima*: Evaluations of releases of *Encarsia luteola* (Hymenoptera: Aphelinidae) and *Delphastus pusillus* (Coleoptera: Coccinellidae). *Environ. Entomol.*, 23, 1346–1353.

Heinz, K.M., Brazzel, J.R., Pickett, C.H., Natwick, E.T., Nelson, J.M. and Parrella, M.P. (1994) Predatory beetle may suppress silverleaf whitefly. *Calif. Agric.*, 48, 35–40.

Hoddle, M.S. and van Driesche, R.G. (1999) Evaluation of inundative releases of *Eretmocerus eremicus* and *Encarsia formosa* Beltsville strain in commercial greenhouses for control of *Bemisia argentifolii* (Hemiptera: Aleyrodidae) on poinsettia stock plants. *J. Econ. Entomol.*, 92, 811–824.

Hoddle, M.S., van Driesche, R.G. and Sanderson, J.P. (1997) Biological control of *Bemisia argentifolii* (Homoptera: Aleyrodidae) on poinsettia with inundative releases of *Encarsia formosa* Beltsville strain (Hymenoptera: Aphelinidae): Can parasitoid reproduction augment inundative releases? *J. Econ. Entomol.*, 90, 910–1240.

Hoffman, J.D. and Ignoffo, C.M. (1974) Growth of *Pteromalus puparum* in a semi-synthetic medium. *Ann. Entomol. Soc. Am.*, 67, 524–525.

Hoffman, J.D., Ignoffo, C.M. and Dickerson, W.A. (1975) *In vitro* rearing of the endoparasitic wasp, *Trichogramma pretiosum*. *Ann. Entomol. Soc. Am.*, 68, 335–336.

Gary W. Elzen, Randy J. Coleman, and Edgar G. King

House, H.C. (1978) An artificial host: encapsulated synthetic medium for *in vitro* oviposition and rearing the endoparasitoid *Itoplectis conquisitor* (Hymenoptera: Ichneumonidae). *J. Econ. Entomol.*, **71**, 331–333.

Howard, L.O. and Fiske, W.F. (1911) The importation into the United States of the parasites of the gypsy moth and the brown tail moth. *U.S. Dept. Agric., Bureau Entomol., Bull. 91.*

Hu, J.S. and Vinson, S.B. (1998) The *in vitro* development from egg to prepupa of *Campoletis sonorensis* (Hymenoptera: Ichneumonidae) in an artificial medium: importance of physical factors. *J. Insect Physiol.*, **44**, 455–462.

Huffaker, C.B. (1974) *Biological Control*, Plenum Press, New York.

Huffaker, C.B. (1977) Augmentation of natural enemies in the People's Republic of China. In R.L. Ridgway and S.B. Vinson (eds.), *Biological Control by Augmentation of Natural Enemies*, Plenum Press, New York, pp. 329–340.

Huffaker, C.B. and Kennett, C.E. (1956) Experimental studies on predation: predation and cyclamen mite populations on strawberries in California. *Hilgardia*, **26**, 191–222.

Huffaker, C.B. and Messenger, P.S. (1976) *Theory and Practice of Biological Control*, Academic Press, New York.

Huffaker, C.B., Rabb, R.L. and Logen, J.A. (1977) Some aspects of population dynamics relative to augmentation of natural enemy action. In R.L. Ridgway and S.B. Vinson (eds.), *Biological Control by Augmentation of Natural Enemies*, Plenum Press, New York, pp. 3–38.

Jaynes, H.A. and Bynum, E.K. (1941) Experiments with *Trichogramma minutum* Riley as a control of the sugarcane borer in Louisiana, *USDA Tech. Bull. 743.*

Jepson, P.C. (1989) *Pesticides and Non-Target Invertebrates*, Intercept, Wimborne, Dorset, UK.

Johnson, M.W. and Tabashnik, B.E. (1999) Enhanced biological control through pesticide selectivity. In T.S. Bellows and T.W. Fisher (eds.), *Handbook of Biological Control*, Academic Press, San Diego, CA, pp. 297–318.

Johnson, W.L., Cross, W.H., McGovern, W.L. and Mitchell, H.C. (1973) Biology of *Heterolaccus grandis* in a laboratory culture and its potential as an introduced parasite of the boll weevil in the United States. *Environ. Entomol.*, **2**, 112–118.

Kanour, W.W. and Burbutis, P. (1984) *Trichogramma nubilale* (Hymenoptera: Trichogrammatidae) field releases in corn and a hypothetical model for control of European corn borer (Lepidoptera: Pyralidae). *J. Econ. Entomol.*, **77**, 103–107.

King, E.G. and Coleman, R.J. (1989) Potential for biological control of *Heliothis* species. *Annu. Rev. Entomol.*, **34**, 53–75.

King, E.G. and Leppla, N.C. (1984) *Advances and Challenges in Insect Rearing*, USDA-ARS, Southern Region, New Orleans, LA.

King, E.G. and Morrison, R.K. (1984) Some systems for production of eight entomophagous arthropods. In E.G. King and N.C. Leppla (eds.), *Advances and Challenges in Insect Rearing*, USDA-ARS, Southern Region, New Orleans, LA, pp. 206–222.

King, E.G., Powell, J.E. and Smith, J.W. (1982) Prospects for utilization of parasites and predators for management of *Heliothis* spp. *Proceedings, Workshop on Heliothis Management*, International Crops Research Institute for the Semi-Arid Topics, Patancheru, A.P., India, pp. 103–122.

King, E.G., Ridgway, R.L. and Hartstack, A.L. (1984) Propagation and release of entomophagous arthropods for control by augmentation. In P.L. Adkisson and Ma Shinjun (eds.), *Proceedings, Chinese Academy of Science–U.S. National Academy of Science Joint Symposium on Biological Control of Insects*, Science Press, Beijing, China.

King, E.G., Hopper, K.R. and Powell, J.E. (1985a) Analysis of systems for biological control of crop arthropod pests in the U.S. by augmentation of predators and parasites. In M.A. Hoy and D.C. Herzog (eds.), *Biological Control in Agricultural IPM Systems*, Academic Press, New York, pp. 201–227.

King, E.G., Bull, D.L., Bouse, L.F. and Phillips, J.R. (1985b) Biological control of bollworm and tobacco budworm in cotton by augmentative releases of *Trichogramma*, Southwest. Entomol., Nov. 8, 1985, Southwest. Entomol. Soc. Press, College Station, TX (suppl.).

King, E.G., Hartley, G.G., Martin, D.F., Smith, J.W., Summers, T.E. and Jackson, R.D. (1979) Production of tachinid *Lixophaga diatraeae* on its natural host, the sugarcane borer, and on an unnatural host, the greater wax moth. USDA, SEA, *Adv. Agric. Tech., South. Ser. No. 3*, 16 pp.

King, E.G., Summy, K.R., Morales-Ramos, J.A. and Coleman, R.J. (1993) Integration of boll weevil biological control by inoculative/augmentative releases of the parasite *Catolaccus grandis* in short-season cotton. In D.J. Herber and D.A. Richter (eds.), *Proc. Beltwide Cotton Conf.*, Cotton Council of America, Memphis, TN, pp. 915–921.

King, E.G., Coleman, R.J., Wood, L., Wendel, L., Greenberg, S., Scott, A.W., Roberson, J. and Hardee, D.D. (1995) Suppression of boll weevil in commercial cotton by augmentative releases of a wasp parasite, *Catolaccus grandis*. In D.A. Richter and J. Armour (eds.), *Addendum to the Proc. Beltwide Cotton Conf.*, Cotton Council of America, Memphis, TN, pp. 26–30.

Knipling, E.F. (1977) The theoretical basis for augmentation of natural enemies. In R.L. Ridgway and S.B. Vinson (eds.), *Biological Control by Augmentation of Natural Enemies*, Plenum Press, New York, pp. 79–124.

Knipling, E.F. (1992) *Principles of Insect Parasitism Analyzed from New Perspectives. Practical Implications for Regulating Insect Populations by Biological Means*, USDA, ARS, Washington, DC.

Kolodny-Hirsch, D.M., Readdon, R.C., Thorpe, K.W. and Raupp, M.J. (1988) Evaluating the impact of sequential releases of *Cotesia melanoscela* (Hymenoptera: Braconidae) on *Lymantria dispar* (Lepidoptera: Lymantriidae). *Environ. Entomol.*, **17**, 403–408.

Laing, J.E. and Hamai, J. (1976) Biological control of insect pests and weeds by imported parasites, predators, and pathogens. In C.B. Huffaker and P.S. Messenger (eds.), *Theory and Practice of Biological Control*, Academic Press, New York, pp. 42–80.

Lawson, D.S., Nyrop, J.P. and Reissig, W.H. (1997) Assays with commercially produced *Trichogramma* (Hymenoptera: Trichogrammatidae) to determine suitability for obliquebanded leafroller (Lepidoptera: Tortricidae) control. *Environ. Entomol.*, **26**, 684–693.

Lawson, F.R., Rabb, R.L., Guthrie, F.E. and Boweny, L.G. (1961) Studies of an integrated control system for hornworms on tobacco. *J. Econ. Entomol.*, **54**, 93–97.

Legner, E.F. and Medved, R.A. (1979) Influence of parasitic Hymenoptera on the regulation of pink bollworm, *Pectinophora gossypiella*, on cotton in the lower Colorado desert. *Environ. Entomol.*, **8**, 922–930.

Letourneau, D.K. and Altieri, M.A. (1999) Environmental management to enhance biological control in agroecosystems. In T.S. Bellows and T.W. Fisher (eds.), *Handbook of Biological Control*, Academic Press, San Diego, CA, pp. 319–354.

Lewis, W.J. and Nordlund, D.A. (1985) Behavior modifying chemicals to enhance natural enemy effectiveness. In M.A. Hoy and D.C. Herzog (eds.), *Biological Control in Agricultural IPM Systems*, Academic Press, New York, pp. 89–100.

Li, L. (1984) Research and utilization of *Trichogramma* in China. In P.L. Adkisson and Ma Shijun (eds.), *Proceedings, Chinese Academy of Science–U.S. National Academy of Science Joint Symposium on Biological Control of Insects*, Science Press, Beijing, China, pp. 204–223.

Luck, R.F., Shepard, B.M. and Kenmore, P.E. (1988) Experimental methods for evaluating arthropod natural enemies. *Annu. Rev. Entomol.*, **33**, 367–391.

Luttrell, R.G., Roush, R.T., Ali, A., Mink, J.S., Reid, M.R. and Snodgrass, G.L. (1987) Pyrethroid resistance in field populations of *Heliothis virescens* (Lepidoptera: Noctuidae) in Mississippi in 1986. *J. Econ. Entomol.*, **80**, 985–989.

Mackauer, M. (1972) Genetic aspects of insect control. *Entomophaga*, **17**, 27–48.

Mackauer, M. (1976) Genetic problems in the production of biological control agents. *Annu. Rev. Entomol.*, **21**, 369–385.

Martin, P.B., Ridgway, R.L. and Schultze, C.E. (1978) Physical and biological evaluation of an encapsulated diet for rearing *Chrysopa carnea*. *Fla. Entomol.*, **61**, 145–152.

McMurtry, J.A., Johnson, H.G. and Scriven, G.T. (1969) Experiments to determine effects of mass release of *Stethorus picipes* on the level of infestations of the avocado brown mite. *J. Econ. Entomol.*, **62**, 1216–1221.

Michelbacher, A.E. and Smith, R.F. (1943) Some natural factors limiting the abundance of the alfalfa butterfly. *Hilgardia*, **15**, 369–397.

Morales-Ramos, J.A., Summy, K.R., Roberson, J.L., Cate, J.R. and King, E.G. (1992) Feasibility of mass rearing *Catolaccus grandis*, a parasitoid of the boll weevil. In D.J. Herber and D.A. Richter (eds.), *Proc. Beltwide Cotton Conf.*, Cotton Council of America, Memphis, TN, pp. 723–726.

Morales-Ramos, J.A., King, E.G. and Summy, K.R. (1993) Magnitude and timing of *Catolaccus grandis* releases against the cotton boll weevil aided by a simulation model. In D.J. Herber and D.A. Richter (eds.), *Proc. Beltwide Cotton Conf.*, Cotton Council of America, Memphis, TN, pp. 915–922.

Morales-Ramos, J.A., Rojas, M.G., Roberson, R.G., King, E.G., Summy, K.R. and Brazzel, J.R. (1994) Suppression of the boll weevil first generation by augmentative releases of *Catolaccus grandis* in Aliceville, Alabama. In D.J. Herber and D.A. Richter (eds.), *Proc. Beltwide Cotton Conf.*, Cotton Council of America, Memphis, TN, pp. 958–964.

Morales-Ramos, J.A., Rojas, M.G. and King, E.G. (1997) Mass propagation of *Catolaccus grandis* in support of large scale area suppression of the boll weevil in South Texas cotton. In P. Dugger and D.A. Richer (eds.), *Proc. Beltwide Cotton Conf.*, Cotton Council of America, Memphis, TN, pp. 199–1200.

Morales-Ramos, J.A., Rojas, M.G., Coleman, R.J. and King, E.G. (1998) Potential use of *in vitro* reared *Catolaccus grandis* (Hymenoptera: Pteromalidae) for biological control of the boll weevil (Coleoptera: Curculionidae). *J. Econ. Entomol.*, **91**, 101–109.

Gary W. Elzen, Randy J. Coleman, and Edgar G. King

Morrison, R.K. and King, E.G. (1977) Mass production of natural enemies. In R.L. Ridgway and S.B. Vinson (eds.), *Biological Control by Augmentation of Natural Enemies*, Plenum Press, New York, pp. 173–217.

Nettles, W.C. Jr. (1982) Contact stimulants from *Heliothis virescens* that influence the behavior of females of the tachinid *Eucelatoria bryani*. *J. Chem. Ecol.*, **8**, 1183–1191.

Nettles, W.C. Jr. and Burks, M.L. (1975) A substance from *Heliothis virescens* larvae stimulating larviposition by females of the tachinid, *Archytas marmoratus*. *J. Insect Physiol.*, **21**, 965–978.

Nettles, W.C. Jr., Wilson, C.M. and Ziser, S.W. (1980) A diet and methods for the *in vitro* rearing of the tachinid, *Eucelatoria* spp. *Ann. Entomol. Soc. Am.*, **73**, 180–184.

Noldus, L.P.J.J. and van Lenteren, J.C. (1983) Kairomonal effects on searching for eggs of *Pieris brassicae, Pieris rapae,* and *Mamestra brassicae* of the parasite *Trichogramma evanescens* Westwood. *Med. Fac. Landbouww. Rijksuniv. Gent.*, **48**, 183–194.

Noldus, L.P.J.J., Lewis, W.J., Tumlinson, J.H. and van Lenteren, J.C. (1988a) Olfactometer and wind tunnel experiments on the role of sex pheromones of noctuid moths in the foraging behaviours of *Trichogramma* spp. *Colloq. INRA*, **43**, 223–238.

Noldus, L.P.J.J., Buser, J.H.M. and Vet, L.E.M. (1988b) Volatile semiochemicals in host-community location by egg parasitoids. *Colloq. INRA*, **48**, 19–20.

Nordlund, D.A., Chelfant, R.B. and Lewis, W.J. (1985) Response of *Trichogramma pretiosum* females to extracts of two plants attacked by *Heliothis zea*. *Agric. Ecosyst. Environ.*, **12**, 127–133.

Nordlund, D.A., Jones, R.L. and Lewis, W.J. (1981a) *Semiochemicals: Their Role in Pest Control*, Wiley, New York.

Nordlund, D.A., Lewis, W.J., Goss, H.R. Jr. and Beevers, M. (1981b) Kairomones and their use for management of entomophagous insects. XII. The stimulatory effects of host eggs and the importance of host egg density to the effective use of kairomones for *Trichogramma pretiosum* Riley. *J. Chem. Ecol.*, **7**, 909–917.

Norton, A.P. and Welter, S.C. (1996) Augmentation of the egg parasitoid *Anaphes iole* (Hymenoptera: Mymaridae) for *Lygus hesperus* (Heteroptera: Miridae) management in strawberries. *Environ. Entomol.*, **25**, 1406–1414.

Oatman, E.R. and Platner, G. (1971) Biological control of the tomato fruitworm, cabbage looper, and hornworms on processing tomatoes in southern California, using mass releases of *Trichogramma pretiosum*. *J. Econ. Entomol.*, **64**, 501–506.

Oatman, E.R. and Platner, G. (1978) Effect of mass releases of *Trichogramma pretiosum* against lepidopterous pests on processing tomatoes in southern California, with notes on host egg population trends. *J. Econ. Entomol.*, **71**, 896–900.

Oatman, E.R. and Platner, G. (1985) Biological control of two avocado pests. *Calif. Agric.*, pp. 21–23.

Oatman, E.R., McMurtry, J.A. and Voth, V. (1968) Suppression of the twospotted spider mite on strawberry with mass releases of *Phytoseiulus persimilis*. *J. Econ. Entomol.*, **61**, 1517–1521.

Oatman, E.R., Gilstrap, F.E. and Voth, V. (1976) Effect of different release rates of *Phytoseiulus persimilis* (Acarina: Phytoseiidae) on the twospotted spider mite on strawberry in southern California. *Entomophaga*, **21**, 269–273.

Oatman, E.R., McMurtry, J.A., Gilstrap, F.E. and Voth, V. (1977) Effect of releases of *Amblyseius californica, Phytoseiulus persimilis,* and *Typhlodromus occidentalis* on the twospotted spider mite on strawberry in southern California. *J. Econ. Entomol.*, **70**, 45–47.

Oatman, E.R., Wyman, J.A., Van Steenwyk, R.A. and Johnson, M.W. (1983) Integrated control of the tomato fruitworm (Lepidoptera: Noctuidae) and other lepidopterous pests on fresh-market tomatoes in southern California. *J. Econ. Entomol.*, **76**, 1363–1369.

Pak, G.A., Noldus, L.P.J.J., van Alebeek, F.A.N. and van Lenteren, J.C. (1988) The use of *Trichogramma* egg parasites in the inundative biological control of lepidopterous pests of cabbage in The Netherlands. *Ecol. Bull.*, **39**, 111–113.

Parker, F.D. (1971) Manipulation of pest populations by manipulating densities of both hosts and parasites through periodic releases. In C.B. Huffaker (ed.), *Biological Control*, Plenum Press, New York, pp. 365–376.

Parrella, M.P., Paine, T.D., Bethke, J.A., Robb, K.L. and Hall, J. (1991) Evaluation of *Encarsia formosa* (Hymenoptera: Aphelinidae) for biological control of sweet potato whitefly (Homoptera: Aleyrodidae) on poinsettia. *Environ. Entomol.*, **20**, 713–719.

Plapp, F.W. Jr. and Bull, D.L. (1989) Modifying chemical control practices to preserve natural enemies. In E.G. King and R.D. Jackson (eds.), *Increasing the Effectiveness of Natural Enemies*. Rekha Printers Publ., New Delhi, India, pp. 537–546.

Powell, J.E. and Hartley, G.G. (1987) Rearing *Microplitis croceipes* (Hymenoptera: Braconidae) and other parasitoids of Noctuidae on multicellular host-rearing trays. *J. Econ. Entomol.*, **80**, 968–971.

Ridgway, R.L. and Vinson, S.B. (1977) *Biological Control by Augmentation of Natural Enemies*, Plenum Press, New York.

Ridgway, R.L., King, E.G. and Carrillo, J.L. (1977) Augmentation of natural enemies for control of plant pests in the Western Hemisphere. In R.L. Ridgway and S.B. Vinson (eds.), *Biological Control by Augmentation of Natural Enemies*, Plenum Press, New York, pp. 379–416.

Roberson, J.L. and Harsh, D.K. (1993) Mechanized production processes to encapsulate boll weevil larvae (*Anthonomus grandis*) for mass production of *Catolaccus grandis*. In D.J. Herber and D.A. Richter (eds.), *Proc. Beltwide Cotton Conf.*, Cotton Council of America, Memphis, TN, pp. 922–923.

Rojas, M.G., Morales-Ramos, J.A. and King, E.G. (1996) *In vitro* rearing of the boll weevil (Coleoptera: Curculionidae) ectoparasitoid *Catolaccus grandis* (Hymenoptera: Pteromalidae) on meridic diets. *J. Econ. Entomol.*, **89**, 1095–1104.

Rojas, M.G., Morales-Ramos, J.A. and King, E.G. (1997) Reduction of cost of mass propagating *Catolaccus grandis* by the use of artificial diet. In P. Dugger and D.A. Richer (eds.), *Proc. Beltwide Cotton Conf*, Cotton Council of America, Memphis, TN, pp. 1197–1199.

Rojas, M.G., Morales-Ramos, J.A. and King, E.G. (1999) Synthetic diet for rearing the hymenopterous ectoparasitoid *Catolaccus grandis*. *U.S. Patent* 5,899,168.

Rude, P. (1984) *Integrated Pest Management for Cotton in the Western Region of the United States*, University of California Press, Oakland, CA.

Sabelis, M.W. and Dicke, M. (1985) Long-range dispersal and searching behavior. In W. Helle and M.W. Sabelis (eds.), *Spider mites: Their Biology, Natural Enemies and Control*, Elsevier Science, Amsterdam, pp. 141–160.

Shands, N.A., Simpson, G.W. and Gordon, C.C. (1972) Insect predators for controlling aphids on potatoes. 5. Numbers of eggs and schedules for introducing them in large field cages. *J. Econ. Entomol.*, **65**, 810–817.

Shands, N., Simpson, G.W. and Simpson, B.A. (1975) Evaluations of field introductions of two insect parasites (Hymenoptera: Braconidae) for controlling potato-infesting aphids. *Environ. Entomol.*, **4**, 499–503.

Shumakov, E.M. (1977) Ecological principles associated with augmentation of natural enemies. In R.L. Ridgway and S.B. Vinson (eds.), *Biological Control by Augmentation of Natural Enemies*, Plenum Press, New York, pp. 39–78.

Simmons, G.S. and Minkenberg, O.P.J.M. (1994) Field-cage evaluation of augmentative biological control of *Bemesia argentifolii* (Homoptera: Aleyrodidiae) in southern California cotton with the parasitoid *Eretmocerus* nr. *californicus* (Hymenoptera: Aphelinidae). *Environ. Entomol.*, **23**, 1552–1557.

Smith, C.N. (1966) *Insect Colonization and Mass Production*, Academic Press, New York.

Snodgrass, G.L. and Scott, W.P. (1988) Tolerance of the tarnished plant bug to dimethoate and acephate in different areas of the Mississippi Delta. In D.J. Herber and D.A. Richter (eds.), *Proc. Beltwide Cotton Conf.*, Cotton Council of America, Memphis, TN, pp. 294–295.

Starks, K.J., Burton, R.L., Teetes, G.L. and Wood, E.A. (1976) Release of parasitoids to control greenbugs on sorghum. *USDA-ARS-S-91*, 12 pp.

Starter, N.H. and Ridgway, R.L. (1977) Economic and social considerations for the utilization of augmentation of natural enemies. In R.L. Ridgway and S.B. Vinson (eds.), *Biological Control by Augmentation of Natural Enemies*, Plenum Press, New York, pp. 431–450.

Stinner, R.E., Ridgway, R.L., Coppedge, J.R., Morrison, R.K. and Dickerson, W.A. (1974) Parasitism of *Heliothis* eggs after field releases of *Trichogramma pretiosum* in cotton. *Environ. Entomol.*, **3**, 492–500.

Summy, K.R., Morales-Ramos, J.A. and King, E.G. (1992) Ecology and potential impact of *Catolaccus grandis* (Burks) on boll weevil infestations in the lower Rio Grande Valley. *Southwest Entomol.*, **17**, 279–288.

Summy, K.R., Morales-Ramos, J.A., King, E.G., Wolfenbarger, D.A., Coleman, R.J., Greenberg, S.M., Scott, A.W. Jr. and French, J.V. (1994) Integration of boll weevil parasite augmentation into the short-season cotton production system of the Lower Rio Grande Valley. In D.J. Herber and D.A. Richter (eds.), *Proc. Beltwide Cotton Conf.*, Cotton Council of America, Memphis, TN, pp. 953–957.

Summy, K.R., Morales-Ramos, J.A. and King, E.G. (1995) Suppression of boll weevil infestations on South Texas cotton by augmentative releases of the exotic parasite *Catolaccus grandis* (Hymenoptera: Pteromalidae). *Biol. Cont.*, **5**, 523–529.

Tabashnik, B.E. and Johnson, M.W. (1999) Evolution of pesticide resistance in natural enemies. In T.S. Bellows and T.W. Fisher (eds.), *Handbook of Biological Control*, Academic Press, San Diego, CA, pp. 673–690.

Thompson, S.N. (1982) *Exeristes roborator:* quantitative determination of *in vitro* larval growth rates in synthetic media with different glucose concentrations. *Exp. Parasitol.*, **54**, 229–234.

Thompson, S.N. (1983) Larval growth of the insect parasite *Brachymeria lasus* reared *in vitro*. *J. Parasitol.*, **69**, 425–427.

Gary W. Elzen, Randy J. Coleman, and Edgar G. King

Thompson, S.N. (1999a) Nutrition and culture of entomophagous insects. *Annu. Rev. Entomol.*, **44**, 561–592.

Thompson, S.N. (1999b) Nutrition of entomophagous insects and other arthropods. In T.S. Bellows and T.W. Fisher (eds.), *Handbook of Biological Control*, Academic Press, San Diego, CA, pp. 594–652.

Thompson, S.N., Bedner, L. and Nadel, H. (1983) Artificial culture of the insect parasite *Pachycrepoides vindemiae*. *Entomol. Exp. Appl.*, **33**, 121–122.

Vanderzant, E.S. (1973) Improvements in the rearing diet for *Chrysopa carnea* and the amino acid requirements for growth. *J. Econ. Entomol.*, **66**, 336–338.

Van der Schaaf, P.A., Kaskens, J.W.M., Kole, M., Noldus, L.P.J. and Pak, G.A. (1984) Experimental releases of two strains of *Trichogramma* spp. against lepidopteran pests in a Brussels sprouts field crop in The Netherlands. *Med. Fac. Landbouww. Rijksuniv. Gent.*, **49**, 803–813.

Van Lenteren, J.C. (1986) Parasitoids in the greenhouse: successes with seasonal inoculative release systems. In J.K. Waage and D.J. Greathead (eds.), *Insect Parasitoids*, Academic Press, New York, pp. 341–374.

Van Lenteren, J.C., Glas, P.C.G. and Smits, P.H. (1982) Evaluation of control capabilities of *Trichogramma* and results of laboratory and field research on *Trichogramma* in the Netherlands. *Les Trichogrammes, Antibes*, Ed. INRA Publ., pp. 257–268.

Vet, L.E.M., van Lenteren, J.C. and Woets, J. (1980) The parasite–host relationship between *Encarsia formosa* (Hymenoptera: Aphelinidae) and *Trialeurodes vaporariorum* (Homoptera: Aleyrodidae). IX. A review of the biological control of the greenhouse whitefly with suggestions for future research. *Z. Angew. Entomol.*, **90**, 26–51.

Vinson, S.B. (1975) Biochemical coevolution between parasitoids and their hosts. In P.W. Price (ed.), *Evolutionary Strategies of Parasitic Insects and Mites*, Plenum Press, New York, pp. 14–48.

Vinson, S.B. (1977) Behavioral chemicals in the augmentation of natural enemies. In R.L. Ridgway and S.B. Vinson (eds.), *Biological Control by Augmentation of Natural Enemies*, Plenum Press, New York, pp. 237–279.

Voronin, K.E. and Grinberg, A.M. (1981) The current status and prospects of *Trichogramma* utilization in the USSR. In J.R. Coulson (ed.), *First Proceedings on Joint American-Soviet Conference on Use of Beneficial Organisms in the Control of Crop Pests*, Entomological Society of America, College Park, MD, pp. 49–51.

Waage, J.K. (1989) The population ecology of pest–pesticide–natural enemy interactions. In P.C. Jepson (ed.), *Pesticides and Non-Target Invertebrates*, Intercept, Wimborne, Dorset, UK, pp. 81–93.

Wackers, F.L., de Groot, I.J.M., Noldus, L.P.J.J. and Hassan, S.A. (1987) Measuring host preference of *Trichogramma* egg parasites: an evaluation of direct and indirect methods. *Med. Fac. Landbouww. Rijksuniv. Gent.*, **52**, 339–348.

Weseloh, R.M. (1984) Behavior of parasites and predators: influences on manipulative strategies and effectiveness. In P.L. Adkisson and Ma Shijun (eds.), *Proceedings, Chinese Academy of Science–U.S. National Academy of Science Joint Symposium on Biological Control of Insects*, Science Press, Beijing, China, pp. 168–180.

Whitten, M.J. and Hoy, M.J. (1999) Genetic improvement and other genetic considerations for improving the efficacy and success rate of biological control. In T.S. Bellows and T.W. Fisher (eds.), *Handbook of Biological Control*, Academic Press, San Diego, pp. 271–296.

Wu, Z., Qin, J., Li, P.X., Chang, Z.P. and Liu, T.M. (1982) Culturing *Trichogramma dendrolimi in vitro* with artificial media devoid of insect material. *Acta Entomol. Sin.*, **25**, 128–135.

Young, J.H. and Willson, L.J. (1984) A model to predict damage reduction to flower buds or fruit by *Heliothis* spp. in the absence or presence of two coleopteran predators. *Southwest Entomol.*, **6**, 33–38.

Zchori-Fein, E., Roush, R.T. and Sanderson, J.P. (1994) Potential for integration of biological and chemical control of greenhouse whitefly (Homoptera: Aleyrodidae) using *Encarsia formosa* (Hymenoptera: Aphelinidae) and abamectin. *Environ. Entomol.*, **23**, 1277.

Developing *Trichogramma* as a Pest Management Tool

4

Linda Thomson[1], David Bennett[1], DeAnn Glenn[2], and Ary Hoffmann[1]
[1]*Centre for Environmental Stress and Adaptation Research, La Trobe University, Bundoora, Victoria, Australia;* [2]*Grape and Wine Development Corporation, Kent Town, South Australia, Australia*

Introduction

Interest in using insect predators and parasitoids to manage pest populations has increased as the undesirable consequences of excessive pesticide use become more evident. In addition to a growing public awareness of the negative impact of pesticides on health and the environment, agricultural chemicals are becoming more expensive and difficult to develop because of increasing resistance to pesticides by key pests (Naranjo, 1993). Some pests are also proving difficult to control economically with pesticides owing to their lifestyle (e.g., Glenn and Hoffmann, 1997).

Parasitoids can have a major impact in natural and agricultural ecosystems where they influence or regulate the population density of many of their hosts (Godfray, 1994). *Trichogramma* and other egg parasitoids are generally part of the local ecosystem and often contribute to the control of lepidopterous pests in the absence of disruptive pesticides. There are many examples of pest control by naturally occurring *Trichogramma* such as the control of *Helicoverpa* eggs in corn in Brazil (De Sa and Parra, 1994) or control of the noctuids *Brusseola fusca* Fuller and *Jesamina calamistis*, and of the pyralids *Chilo partellus orichalociliellus* and *Eladana saccharina* Walker, which attack maize in east Africa (Bonhof *et al.*, 1997). Control of the cranberry fruitworm, *Acrobasis vaccinii* Riley in the USA (Simser, 1994), and in Canada where *Trichogramma* nr. *sibericum* and *Trichogramma minutum* Riley attack the blackheaded fireworm, *Rhopobota naevana* Riley, in cranberries are excellent examples of such control.

Naturally occurring predators and parasitoids are often not present in sufficient numbers at the right time to keep pest species within an economically sustainable limit. *Trichogramma* release programs can be used to overcome these limitations. There are two ways to use *Trichogramma* release in pest control: "inoculative" releases to maintain and augment an existing population, or "inundative" releases to introduce large numbers of insectary-reared *Trichogramma* to coincide with maximum host presence. Both approaches aim to increase *Trichogramma* parasitism of the pest to reduce crop losses.

Trichogramma have been used in inundative releases more than any other natural enemy (Stinner, 1977). Current situations where *Trichogramma* are used to control lepidopterous pests include grapes (Glenn and Hoffmann, 1997), tomatoes in greenhouses (Shipp and Wang, 1998) and tomatoes in the field (Consoli *et al.*, 1998), rice (Bentur *et al.*, 1994), apples (McDougall and Mills, 1997), and sugar cane (Greenberg *et al.*, 1998a). *Trichogramma* are also used against *Helicoverpa armigera* (Hubner) on a variety of crops in India (Romeis and Shanower, 1996) and on sweet corn in Australia (Scholz *et al.*, 1998). *Trichogramma* are an effective biological control agent against the European corn borer, *Ostrinia nubilalis* Hubner (Lepidoptera: Pyralidae), throughout Europe (Chihrane *et al.*, 1993; Hawlitzky *et al.*, 1994; Mertz *et al.*, 1995; Chihrane and Lauge, 1996; Gunie and Lauge, 1997), Canada (Yu and Byers, 1994), and North America (Andow *et al.*, 1995). They are used in nonfood crops such as cotton (Naranjo, 1993) and to provide foliage protection in forests (Bai *et al.*, 1995). *Trichogramma* are even used against lepidoptera in stored grain, where *Trichogramma evanescans* Westwood and *T. embryophagum* Hartig attack *Ephestia kuehniella* Zeller and *E. elutella* Hubner (Scholler *et al.*, 1996).

Despite the widespread use of *Trichogramma*, there are relatively few cases where the successful control of a pest can be unequivocally ascribed to releases of *Trichogramma*. There are many documented failures of *Trichogramma* releases, despite a few notable successes (Twine and Lloyd, 1982; Bigler and Brunetti, 1986; Smith *et al.*, 1987; Newton and Odendaal, 1990; Li, 1994).

The variable success of *Trichogramma* field releases can be attributed to a number of factors (Smith, 1996). These include pesticide applications, weather conditions at the time of release, lack of host parasitoid synchrony, and the quality of the release material. This last area is complex and encompasses diverse issues such as technical aspects related to rearing, storage, and shipping of the release material as well as the characteristics of the *Trichogramma* chosen for release programs.

In this chapter we focus on quality issues and consider three main questions. First, how do we ensure that the *Trichogramma* strain or species is suited to release conditions? This requires effective screening of genetic variability. Once suitable strains are identified, mass-reared populations need to be monitored to ensure that undesirable characteristics are not selected. Second, which environmental factors can influence quality and how do we manipulate these factors to ensure that the quality of released wasps is maximized? Most environmental factors relate to the nature of the rearing host, which can influence the size and behavior of the adult wasps. Finally, what release conditions influence the success of *Trichogramma*? Releases should ideally be integrated into all aspects of the target pest environment using a systems approach (Lewis *et al.*, 1997). At the end of this chapter we provide a case study of our own

Linda Thomson, David Bennett, DeAnn Glenn, and Ary Hoffmann

work aimed at the development of *Trichogramma* for the control of lightbrown apple moth (Tortricidae: *Epiphyas postvittana* Walker) in vineyards in Australia.

Quality issues: genetic and environmental components

Phenotypic variation can have genetic and environmental components. Environmental variation is often mediated via the size of the host egg, which in turn affects the size of the *Trichogramma* (e.g., Glenn and Hoffmann, 1997; Bennett and Hoffmann, 1998). In addition there are genetic factors that can act independently of size. For example, Bigler *et al.* (1988) and Dutton *et al.* (1996) found that walking speed was affected by genetic factors independent of size, while Ashley *et al.* (1974) found genetic variation in locomotion and heat tolerance.

Genetic quality: identification and maintenance of appropriate strains

Successful commercial release programs require *Trichogramma* that are effective against target hosts under field conditions. The overall release quality of *Trichogramma* will be influenced by many traits. Identifying which traits are related to optimal release quality is a major problem when screening for suitable strains. Which trait(s) will be important depends on the specific field situation in which the parasitoids are being used. Once an effective strain has been identified, quality needs to be maintained, avoiding selection for undesirable traits in mass-rearing conditions.

There are many examples where *Trichogramma* species and strains of the same species differ in traits that may be related to field performance. For instance, the ability of foraging females to discover hosts has been found to vary among strains. Foraging ability is likely to be related to the surface area that females prospect per unit time (Bigler *et al.*, 1993) and to walking speed (Bigler *et al.*, 1988, 1993; Dutton *et al.*, 1996; Wajnberg and Collazo, 1998). Strain variation has also been found for longevity (Bigler *et al.*, 1993; Dutton *et al.*, 1996), emergence rates (Frei and Bigler, 1993; Dutton *et al.*, 1996), sex ratio (Bigler *et al.*, 1993), and fecundity on both natural and factitious hosts (Bigler *et al.*, 1993; Frei and Bigler, 1993; Dutton *et al.*, 1996).

Trichogramma strains can also differ in the extent to which they respond to environmental change. Variation has been found between species and strains for their ability to survive and parasitize under different temperatures (Pak and van Heiningen, 1985; Pavlik, 1990). The variation in *Trichogramma* strains for cold tolerance and adaptability to low temperatures makes this a useful criterion for evaluation of candidate strains for inundative releases in cold conditions (Voegele *et al.*, 1988). Flight propensity (Forsse *et al.*, 1992) and rates of travel (Boldt, 1974) are also affected by temperature and may represent useful traits for screening.

While it is clear that variation exists among *Trichogramma* strains and species for a number of traits that may contribute to the success of field releases, there are three questions that need to be addressed when assessing the potential impact of this variability.

1. What is the relative contribution of genetic versus environmental factors to the phenotypic variation? Identifying strain differences within one generation represents a first step in a genetic analysis, but provides no information whether variation persists across generations, particularly if environmental conditions change. Genetic studies need to encompass more than one generation and should ideally consider the heritability of traits across different environments. Examples of genetic studies on parasitoids encompassing more than one generation include Wajnberg and Colazzo (1998), Schumacher *et al.* (1997), and Bennett and Hoffmann (1998). Unfortunately, *Trichogramma* studies generally do not address whether strain differences persist across different environmental conditions.

2. What are the effects of mass rearing on phenotypic variation? Once useful strains have been identified, phenotypes can still change owing to commercial conditions. Mass-rearing facilities usually have constant temperature and regulated periods of dark and light, with high parasitoid host densities and limited need for host search-ing behaviors such as walking or flying. The presence of genetic variation within species means that *Trichogramma* may readily adapt to conditions used in mass-rearing facilities (Sorati *et al.*, 1996). This appears to be one of the major prob-lems encountered in quality control of *Trichogramma* for mass release (van Bergelijk *et al.*, 1989), because as strains adapt they may become ineffective as biological control agents (Sorati *et al.*, 1996), although this is not necessarily always the case (Ashley *et al.*, 1973). Inbreeding effects may also cause a decline in quality in commercial cultures. However, as *Trichogramma* are haplodiploid, major inbreed-ing effects are unlikely. Any deleterious alleles that arise (most of which will be recessive) will rapidly be selected against in the haploid males. Sorati *et al.* (1996) suggest that the absence of inbreeding effects in *Trichogramma* could be used in the maintenance of mass-reared colonies. This method would require a colony with limited variability. If the strains selected for mass release could be intensively inbred to fix useful genes and minimize their loss during adaptation to artificial environments, desirable traits could be maintained. Thus, inbreeding may be a useful tool in minimizing the effects of laboratory adaptation (Figure 4.1).

3. Which traits are relevant for field success? There is likely to be an association between locomotion and parasitism in the field. Variation in travel speed has been used to estimate capacity for host location and the efficiency of *T. maidis* Pintureau & Voegele strains for inundative biological control programs (Bigler *et al.*, 1988). There have also been attempts to combine quality parameters in devising an index for quality. In particular, Dutton *et al.* (1996) measured four quality parameters: walking speed, lifespan, and fecundity on the natural host and on the factitious host. A quality index based on three of these traits (walking speed and the two fecundity measures) was calculated and compared for high-quality and low-quality populations of *T. brassicae* Bezdento. Fecundity on the factitious host was found to be a better predictor of success in the field than the quality index.

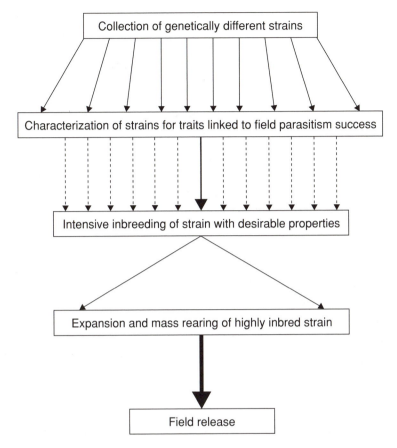

Figure 4.1 Use of inbreeding to develop *Trichogramma* strains for mass release with minimal adaptation to mass-rearing conditions. Strains are first characterized to identify a strain with desirable characteristics. This strain is then bred to remove variability within the strain that would facilitate adaptation to the mass-rearing conditions. The inbred strain is built up to large numbers and used for releases.

Environmental effects: rearing host

To rear *Trichogramma* on a commercial scale, it is necessary to use a factitious rearing host, such as *E. kuehniella* or *Sitotroga cerealella* Olivier, rather than the natural or target host. The choice of factitious host is often dictated by the ease of rearing and not necessarily by any factors related to the likely success of the wasps being produced. Factitious hosts are selected for the simplicity of their mass production, mechanization of rearing processes, and cost of production compared to that of using the target pest (Greenberg *et al.*, 1998b).

There are several potential effects of rearing parasitoid biocontrol agents on a nontarget host. The rearing host can affect qualities such as development time (Bai *et al.*, 1995), travel speed (Boldt, 1974), longevity, percent emergence, and sex ratio

(Corrigan and Laing, 1994). The size of the factitious host egg could alter the size of the emerged wasp, which could have effects on wasp attributes such as target host acceptance and fecundity. Often *Trichogramma* reared from one host will be used to target several pests, introducing many potential complications, as the same wasps may not be suitable across several different hosts.

There appear to be only minor changes in acceptance of the target host when *Trichogramma* are reared on a factitious host. Bjorksten and Hoffmann (1995) found that oviposition experience had a stronger effect on host preference than preadult experience (learning through development in the rearing host). Bjorksten and Hoffmann (1998) found experience effects in *T.* nr. *brassicae* due to rearing host and oviposition by females, but these effects only influenced the likelihood of parasitism in low-ranked hosts and not in high-ranked hosts that would comprise the target pests in a mass release program. In *T. carverae* Oatman and Pinto there was also no difference in acceptance of the target species, *Epiphyas postvittana*, when wasps were reared on the factitious host *S. cerealella* compared to the target host (Glenn and Hoffmann, 1997).

There is evidence that the rearing host can cause variation in size of emerged wasps (Bai *et al.*, 1995). *T. carverae* reared on *E. postvittana* eggs are significantly larger than those reared on *S. cerealella* (Glenn and Hoffmann, 1997), and the size of mass-produced *T. pretiosum* Riley and *T. minutum* adult females is dependent on the size of the rearing host egg on which the insect develops (Greenberg *et al.*, 1998c). But does size matter? There are reports of increased size being associated with increased fitness in parasitoids, measured as success in locating hosts (Visser, 1994; Kazmer and Luck, 1995; West *et al.*, 1996; Bennett and Hoffmann, 1998) and increased fecundity (Bai *et al.*, 1995; Greenberg *et al.*, 1998c). This, however, is not always the case. Bigler *et al.* (1993) found that wasps emerged from *E. kuehniella* were bigger than those emerged from *S. cerealella*, but there was no perceived quality difference. Pavlik (1993) found no significant relationship in *Trichogramma* between hind tibia length and egg complement, fecundity, longevity, or locomotor activity.

Because results may not be consistent across studies, it is necessary to examine factors influencing field performance in chosen species/strains. For instance, while size may be a good indicator of success of *T. carverae* released in grapevines (Bennett and Hoffmann, 1998 — see below), this may not be true of other species/systems. Fecundity on the factitious host may be species/strain-specific, so it is necessary to compare fecundity on the rearing host to that on the target host before a comprehensive cost–benefit analysis of a mass rearing-release system can be made (Corrigan and Laing, 1994; Greenberg *et al.*, 1998c).

Other environmental effects

Apart from host effects, other environmental factors may also impact on wasp quality. In particular, wasps can become acclimated to particular conditions. Mass production facilities rear the wasps at constant temperature (usually 25°C). It is possible that wasps reared under constant conditions in insectaries are of low quality because they fail to survive extreme temperature fluctuations encountered in the field (Scott *et al.*, 1997). This may reduce the effectiveness of releases in cold weather. Scott *et al.*

Linda Thomson, David Bennett, DeAnn Glenn, and Ary Hoffmann

(1997) reared *T. carverae* at three temperatures (14, 25, and 30°C) and found that only wasps reared at 14°C parasitized eggs at that temperature. The substantial reduction in parasitism of the sugarcane borer, *Diatraea saacharalis* Fabricius, by *T. galloi* in the winter months may be related to a thermal alternation shock, as the wasps are mass-reared at constant temperatures (Consoli and Parra, 1995). Turning to another parasitoid, Numata (1993) found diapause adults of the parasitoid wasp *Ooencyrtus nezarae* Ishii reared at 25°C to be more resistant to low temperature than nondiapause adults reared at 25°C.

Responses to high temperature can also be improved via acclimation and may be useful in improving the quality of beneficial insects (Huey and Berrigan, 1996). There is some experimental evidence that heat resistance can be increased by acclimation. Scott *et al.* (1997) found that heat hardening *T. carverae* adult wasps at 33 or 35°C for 1–2 hours increased survivorship at 40°C. *T. brassicae* often experience temperatures as high as 44°C during release in sweet corn (Chihrane and Lauge, 1996). Maisonhaute (1999) found that a heat shock of 35 and 44°C in the laboratory led to the induction of heat shock proteins. At 44°C it also increased pupal mortality, decreased fecundity, and increased male sterility, whereas a 35°C pretreatment had a protective effect when a subsequent 44°C shock was applied.

Unfortunately, any benefits of acclimation could be offset by fitness costs of the acclimation process (Hoffmann, 1995; Huey and Berrigan, 1996). Gunie and Lauge (1997) found that temperature shocks affected the emergence rate and fecundity of females. Even a short low-amplitude shock of 32°C to *T. brassicae* had a strong negative effect on emergence rate and fecundity under benign conditions. While Scott *et al.* (1997) found an increase in survival after developmental hardening, they also recorded decreases in adult longevity and parasitism rate. Beneficial acclimation (acclimation to one temperature that confers a performance advantage at that temperature) is not a necessary outcome of developmental acclimation (Huey *et al.*, 1999). Acclimation seems a promising procedure for releases when temperature extremes are unavoidable, but specific research is needed to determine the optimum temperature, stage of development, and length of exposure for different species/strains to improve performance at high temperatures while minimizing the costs. For instance, acclimation of *T.* nr. *brassicae* at the pupal stage for two hours per day led to increased survival of 40°C heat shock without any associated decrease in parasitism (Hoffmann and Hewa-Kapuge, 2000). While in theory it is possible that acclimation could improve field release success in certain circumstances, field assessments of the costs and benefits of such procedures are essential.

Quality control

To ensure that a high-quality product is delivered to the grower, optimum conditions for rearing, storage, and shipment are imperative. It is important that producers become aware of the negative effects of poor handling of *Trichogramma* (Dutton and Bigler, 1995; Bigler, 1994). Bigler *et al.* (1993) tested the quality of commercially available *Trichogramma* and concluded that more elaborate product control systems were necessary to increase reliability of the product.

Product quality is recognized as one of the most important reasons for failure of the biological control agent *T. chilonis* Ishii against *H. armigera* (Romeis *et al.*, 1998). O'Neil *et al.* (1998) evaluated the quality of four commercially available natural enemies including *T. pretiosum*. The postshipment quality from 10 companies was assessed for emergence rates, sex ratio, survivorship, species identity, reproduction, and parasitism. Considerable differences in the number received, survivorship, and emergence rates were found. Field studies using commercially available *T. brassicae* against *O. nubilalis* in sweet corn differed in emergence profiles in two different years (Mertz *et al.*, 1995). When several commercially available species of *Trichogramma* used against *Plutella xylostella* Linnaeus were compared, inconsistent responses were observed within most of the products, indicating potential problems with quality control (Vasquez *et al.*, 1997).

The insects need to be reared on an industrial scale, and cold storage makes possible the management of large quantities of living material for intensive periods of use. It is necessary for rearing facilities to have reliable systems for storing eggs, pupae, or adults (Chang *et al.*, 1996). Diapause and quiescence constitute major physiological adaptations for sustaining survival during environmental extremes and both adaptations can have practical applications in storage during mass rearing (Zaslavski and Umarova, 1990; Chang *et al.*, 1996).

Unfortunately, cold storage may impact on the success or efficiency of the organisms to be released. These effects include reduced fecundity (Jalali and Singh, 1992; Frei and Bigler, 1993; Hutchinson and Bale, 1994; Chang *et al.*, 1996), poor flight activity (Dutton and Bigler, 1995), reduced longevity (Jalali and Singh, 1992; Hutchinson and Bale, 1994), and reduced emergence rate (Jalali and Singh, 1992; Cerutti and Bigler, 1995).

Release and integration issues

After selection of the optimum species or strain for the target host and ensuring the best product possible after storage and shipment to the point of release, release conditions need to be considered to ensure maximum parasitism. The grower has some control in ensuring host–parasitoid synchrony, avoiding contact with harmful pesticides, controlling the stage of development at the time of release, providing protection from predation, and releasing in optimal weather conditions.

Host parasitoid synchrony

Host availability is the key to realizing the highest possible level of parasitism. For successful augmentative releases, monitoring of the pest species is essential to ensure application of the *Trichogramma* when host eggs are available (Hassan, 1994). Large numbers of released organisms need to be synchronized closely with the start of oviposition in the pest (Smith, 1994). Monitoring of the target host must be continued through the growing season of the crop and further inundative applications need to be made when appropriate. It is also necessary to ensure that target eggs in the area at the time of release will be acceptable to the released *Trichogramma*. A high rate of parasitism will be achieved only if the wasps reach the host eggs when they are susceptible to parasitism (Glenn and Hoffmann, 1997). Reports vary as to the acceptability of host eggs of different ages to *Trichogramma* species. Some eggs are acceptable for most of their development time (Glenn and Hoffmann, 1997), but other studies show that the age of host eggs will affect performance (Miura and Kobayashi, 1998; Shipp and Wang, 1998). Clearly there is a need to examine in detail the species of *Trichogramma* to be used and the acceptability of the target host eggs for that species.

Pesticides

Whether attempts to improve levels of *Trichogramma* parasitism are by inoculative or inundative methods, there is a high potential for failure if there is associated pesticide use. *Trichogramma* is generally sensitive to pesticides (Bartlett, 1963; Franz *et al.*, 1980). Chemicals can be immediately toxic and cause death (contact toxicity) and this effect can persist (toxicity of dried residue) (Hassan *et al.*, 1987, 1988, 1994, 1998). Contact with pesticides at the less susceptible life stage (parasitoids within hosts) can cause prolonged development time of immature stages and reduced rates of emergence, fecundity, parasitism capacity, adult longevity, and mating likelihood (Franz *et al.*, 1980; Hassan *et al.*, 1994; Schmuck *et al.*, 1996; Consoli *et al.*, 1998).

Exposure of *Trichogramma* adults to pesticides may also interfere with mating. Delpuech *et al.* (1999) found that pyrethroids and chlorpyrifos at sublethal doses interfered with sex pheromone communication. Male *T. brassicae* treated with a very low dose of the insecticide deltamethrin failed to respond to females, and the response of untreated males to treated females was also significantly decreased. In *Trichogramma*, like other insects, sex pheromonal communication probably involves nervous transmission for both the reception and emission of the pheromone. Even if insecticides do not provoke mortality, sublethal doses may be a threat for species and population equilibria by modifying physiological or behavioral parameters such as pheromonal communication. This is especially true for species with strong interaction such as parasitoids and hosts (Delpuech *et al.*, 1998a,b).

There is a changing perception of pesticide use as resistance and the effects of chemicals are recognized, leading to ways to minimize chemical use and combine pesticide use with protection and introduction of beneficials in integrated pest management (IPM) programs. As it becomes more apparent that biological control is an important component of pest management, there has been a change in consumer attitude that is leading to the development of suitable new pesticides. Research is currently identifying pesticides suitable for use in conjunction with biological control, as it is clearly preferable to use chemicals that have minimum toxicity to beneficial organisms (Franz *et al.*, 1980). To determine which chemicals will be most suitable for inclusion in an IPM program, insecticides must be tested for their effect on beneficials. A battery of standardized laboratory and semi-field-test protocols based on lethal and sublethal effects as evaluation criteria for the side effects of pesticides on beneficial organisms have been developed by the IOBC/WPRS working group "Pesticides and Beneficial Organisms" (Hassan *et al.*, 1987, 1988, 1994, 1998). The results showed that preparations greatly differ in initial toxicity as well as persistence. Even where the use of toxic chemicals is difficult to avoid, research into persistence could suggest the optimal separation of spray application and *Trichogramma* releases to minimize toxic effects of the pesticide.

Reducing pesticide resistance and replacing insecticide use with inoculative or inundative releases of *Trichogramma* is complicated by the fact that each crop is sprayed with many chemicals. For this reason, rather than approach biological control as a replacement for chemical control, a more conservative IPM approach is often used. In such a program efforts are made to reduce the use of pesticides, to test and use pesticides less toxic to beneficials, and to remove releases of beneficials as far as possible from spray applications. In fact, several researchers report higher parasitoid activity and better control of pests without the use of pesticides (Simser, 1994; Sundaram *et al.*, 1994; Scholz *et al.*, 1998). For instance, in *H. armigera* on sweet corn in Australia, an application of deltamethrin actually reduced the action of a large natural population of *T. pretiosum* wasps and resulted in higher larval infestation and significantly more cob damage (Scholz *et al.*, 1998).

Weather conditions at the time of release

Weather is one of the most important factors influencing the performance of biological control agents (Naranjo, 1993; Wang *et al.*, 1997; Shipp and Wang, 1998). Climatic factors, particularly temperature extremes, must be considered in any project involving pest management with beneficial insects. In India several species of *Trichogramma* have gained importance against different lepidopteran pests, but the performance of these parasitoids under some climatic conditions seems erratic. The high temperature prevailing in the subcontinent is suspected to be responsible (Ramesh and Baskaran, 1996). *T. galloi* is the most common egg parasitoid of the sugarcane borer *D. saccharalis* in Sao Paulo, Brazil. Field releases indicate a substantial reduction in parasitization capacity of *T. galloi* during the winter months (Consoli and Parra, 1995). The likely weather conditions at the precise time of release also need to be considered, particularly where *Trichogramma* may encounter daily extreme temperatures. In desert cotton growing areas of California and Arizona where *Trichogramma* are released to control pink bollworm, *Pectinophora gossypiella* Saunders, ambient air temperatures frequently exceed 40°C (Naranjo, 1993). *T. brassicae* released to control European corn borer, *O. nubilalis* Hubner, in southeast France may be exposed to temperatures as high as 44°C during dispersal in enclosed cardboard capsules (Chihrane and Lauge, 1996).

The number of degree-hours above a critical minimum can be the most important single variable in realized fecundity and host acceptance in field trials (Bourchier and Smith, 1996; Dutton *et al.*, 1996).

Developing *Trichogramma* as a biological control agent for the control of lightbrown apple moth in Australian vineyards: a case study

The *Trichogramma* industry in Australia has only recently become established. The Angoumois grain moth, *S. cerealella* was first introduced as a factitious host for *Trichogramma* rearing in the 1970s by Grimm and Lawrence (1975) and McLaren and Rye (1983). Despite the development of mass rearing, taxonomic collections of Trichogrammatidae in Australia have not been extensive.

Lightbrown apple moth (*E. postvittana*, LBAM) in wine grapes was considered to be a particularly good candidate for control by inundative release of *Trichogramma* for a number of reasons. LBAM larvae can be difficult to control with insecticides as they are protected in webbed leaf rolls or grape bunches, which limit contact with insecticide sprays, especially in dense vine foliage and tight grape bunches late in the season. Wine grape growers are keen to reduce the risk of pesticide residues in wines and are very supportive of the development of biological control systems for grapes. In addition, LBAM resistance to pesticides has been documented in New Zealand (Suckling *et al.*, 1984). LBAM are relatively vulnerable to egg parasitism as egg masses of 20–50 eggs are laid on the upper surface of basal leaves in grapevines and egg hatch occurs after 6–21 days depending on temperatures (Danthanarayana, 1975). Moths tend not to be migratory (Danthanarayana, 1976) and prediction of peak oviposition times is possible using day-degree models and/or moth trapping with port wine lures and/or pheromone traps and egg monitoring (Madge and Stirrat, 1991). Unlike for many other high-value horticultural crops, cosmetic damage in wine grapes is not a major issue, so damage tolerance is higher than in some other cropping systems. The success of *Trichogramma* releases can be assessed by monitoring eggs so that further control measures such as *Bacillus thuringiensis* or other sprays can be used if necessary.

The aim of this biocontrol program was to develop the use of *Trichogramma* as an alternative control strategy for LBAM in wine grapes. Potential *Trichogramma* species were identified and tested under laboratory conditions before assessing methods for improving release strategies in field conditions. Then the focus was on ways of maintaining the quality of wasps, looking specifically at wasp size and asymmetry and also at ways of increasing the tolerance of wasps to temperature extremes. Current research includes a study of the impact of pesticides on *Trichogramma* and an investigation into patterns of oogenesis under conditions of host deprivation.

Identifying effective endemic species

Carver (1978) described a new *Trichogramma* subgenus, *Trichogrammanza*, that is native to Australia. The separation between these subgenera is based on an extra funicular segment in the male antenna. Two species of *Trichogrammanza* have been identified to date, including *Trichogramma (Trichogrammanza) funiculatum* Carver and *Trichogramma (Trichogrammanza) carverae* Oatman & Pinto.

Collections were made from vineyards over several years throughout southeastern Australia and morphological features commonly used to distinguish *Trichogramma* were examined. Male antennal segments were used to separate Trichogrammatoidea, Trichogramma (*Trichogramma*) and *Trichogramma (Trichogrammanza)* (Carver, 1978; Oatman and Pinto, 1987; Glenn *et al.*, 1997). The length and width of the genital capsule, dorsal aperture, length of the parameres, and distance from the tip of the intervolsellar process to the tip of the parameres were measured to distinguish between representative *Trichogrammanza* specimens. In this case, measurements from the tip of the intervolsellar process to the tip of the parameres were useful for separating *T. carverae* and *T. funiculatum* Carver (Glenn *et al.*, 1997). Strains within the subgenus *Trichogrammanza* that could not be confidently identified morphologically to species were investigated further. Collections were separated according to color markings and their mating incompatibility was examined by crossing different strains. Development times of strains were assessed at different temperatures and patterns of host acceptance were compared. Finally, electrophoretic markers were used to distinguish strains genetically. In all, five species of *Trichogramma* were identified from LBAM egg masses with these various approaches (Glenn *et al.*, 1997). These were *T. carverae*, *T. funiculatum*, *Trichogramma (Trichogrammanza)* sp. *x*, *Trichogramma* nr. *brassicae* (also known as *Trichogramma* nr. *ivelae*), and *Trichogrammatoidea* nr. *bactrae*.

In Australia there is one insectary that rears *Trichogramma* and only on *Sitotroga* eggs. It was, therefore, essential that the species chosen could be reared on this factitious host. *Trichogramma* were collected in the field from naturally parasitized LBAM egg masses. After 2–3 generations on LBAM eggs, wasps of the different species were transferred to *Sitotroga* eggs to assess their potential for mass rearing. There was variation between strains for acceptance of *Sitotroga* when eggs were presented to wasps from the field collections. Further screening and selection was directed toward *T. carverae* when it became clear that mass rearing of *T. funiculatum* and *T.* sp. *x* was not possible on *Sitotroga*, and *T.* nr *brassicae* did not readily accept LBAM egg masses. Additional experiments showed that female *T. carverae* reared on *Sitotroga* had lower fecundity than those reared on LBAM, presumably as a result of their smaller size. However, neither longevity, nor acceptance of LBAM eggs were reduced after rearing *T. carverae* on *Sitotroga* compared to LBAM, suggesting that rearing on the factitious host does not seriously compromise the use of *T. carverae* as a biocontrol agent. These findings, therefore, suggested that *T. carverae* was a good candidate for commercial releases.

T. carverae were released at various rates in the field and the success was monitored with sentinel cards consisting of one green LBAM egg mass of 30–50 eggs. Parasitism rates exceeded 80 percent for release rates of 100–150 000 wasps per ha. These results indicate that high parasitism levels can be achieved with *T. carverae* in vineyards. However, other releases by different researchers have produced more variable results, with parasitism rates in some cases being only around 20–30 percent. This raises the issues of why the success of this endemic species has been variable and the need to overcome potential problems.

Linda Thomson, David Bennett, DeAnn Glenn, and Ary Hoffmann

Identifying high-quality wasps

The fitness of the strain may suffer from being mass-reared under artificial conditions. To counter this, an attempt was made to identify traits important in field quality. Fourteen strains were combined to find a heterogeneous *T. carverae* colony. Two field trials were conducted to identify females that were able to search for and locate eggs of the target host LBAM in grapevines. Several fitness components such as dispersal ability, orientation mechanisms and survival rates are likely to be important in this process.

Prior to release, a sample of wasps emerging from the parasitized *Sitotroga* eggs was collected. Female wasps found ovipositing on the eggs after release were also collected. The size and fluctuating asymmetry distributions were then compared between emergence females (i.e., before selection) and females successful in locating oviposition sites consisting of LBAM egg masses on grapevines. In addition, parent–offspring regression analyses were undertaken to examine heritable variation in size and fluctuating asymmetry measures.

Five metric traits were used for the size measures, including forewing length and width, hind wing length, head width, and hind tibia length (HTL). All correlations among size traits were significant and positive. Further analyses were undertaken on HTL, which has been used in previous studies as an estimate of body size (Pavlik, 1993). Distribution data for HTL indicate that ovipositing females were larger than those from the emergence sample (Figure 4.2). Wasps with intermediate sizes tend to have the lowest fitness, while there is an increase in fitness as HTL exceeds 0.14 mm. The rate at which fitness increases with size levels off at higher values, but there is no evidence for a plateau. Mother–daughter regressions were undertaken on the five traits estimating body size, with none differing from zero. An additive genetic

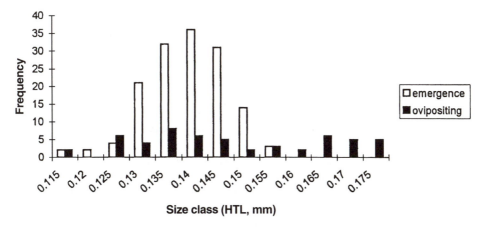

Figure 4.2 Distribution of hind tibia length (in mm) in *T. carverae* wasps from ovipositing and emergence samples. Wasps in ovipositing samples were collected from egg masses in the field. Those from emergence samples were a random sample of the wasps that were released.

component for size was, therefore, not detectable in these samples, suggesting that size variation is environmental.

Two of the metric traits (forewing length and forewing width) used for size studies were measured on both sides for use in the asymmetry analysis. In addition, four meristic traits on the forewings were identified. One count of trichiae and three counts of fringe setae on the forewing were chosen for this study. The asymmetry of each trait was measured as the signed (R–L) difference. Fluctuating asymmetry is the absolute value of this measure. The repeatability of asymmetry was initially tested following the approach outlined in Palmer (1994). A composite asymmetry score for each wasp was obtained by adding fluctuating asymmetry values across several traits.

While the composite index did not differ between samples, there was a difference in the distribution of wing length asymmetry between the emergence and ovipositing samples, even when corrected for multiple comparisons, indicating that low asymmetries are associated with high fitness. These findings suggest that only a small fraction of the released wasps had a high fitness. For instance, wasps with an HTL of 0.16 mm or greater had the highest fitness but would have comprised only 0.04 percent of the release sample. Yet these wasps comprised 27 percent of the ovipositing sample. Wasps reaching oviposition sites are, therefore, an extremely select group of those released. Similar but less spectacular results were obtained for wing asymmetry: Wasps from the lowest asymmetry class comprised 6 percent of the population yet constituted 21 percent of the sample reaching the oviposition sites.

Trichogramma are small and adults can have relatively short lifespans, particularly in the field. Wasps may be exposed to extremes of temperature during this time and it would be useful if wasp survival could be improved by acclimation, without costs such as decreases in fecundity. Experiments were undertaken to determine whether beneficial acclimation occurs as a consequence of heat hardening during development and whether the acclimation treatments influence fecundity or longevity of adult wasps. Wasps reared at 25°C were heat hardened at 33°C for varying lengths of time. Acclimation at 33°C for both the pupal stage (8 hours on day 10) and adult stage (30 minutes on day of emergence) increased survival of wasps subjected to a 40°C heat stress, without a cost to fecundity. This benefit was demonstrated under laboratory and field conditions (Thomson *et al.*, 2001).

Integration issues

The optimal vision of inundative releases of *Trichogramma* replacing insecticides is complicated by the fact that each crop is sprayed with many chemicals. Accordingly, a mechanism to avoid the need to spray with chemicals to control LBAM cannot be implemented if the release organism is susceptible to another chemical applied to the crop for unrelated reasons. It is therefore necessary to determine the effects of chemicals that may be used in the vineyard where *Trichogramma* are to be released. For instance, sulfur has been applied to grapevines for control of powdery mildew and several species of mites for more than 100 years. Control of these pests and diseases

Linda Thomson, David Bennett, DeAnn Glenn, and Ary Hoffmann

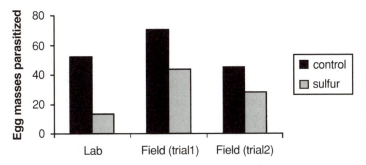

Figure 4.3 Parasitism of sulfur-sprayed and control egg masses in the laboratory and in the field expressed as percent of egg masses accepted.

is regarded as essential and using sulfur has several advantages: it is inexpensive and effective, and has a multisite action that inhibits development of fungicide-resistant strains.

In the laboratory, mortality of *T. carverae* was assessed after direct contact with the pesticide and the effects on persistence and performance after contact with residue were also considered. Mortality and performance of adults were tested in adults emerging from sprayed eggs to test the effects on the less-susceptible stage (parasite within host). In addition, acceptance of eggs sprayed with sulfur was compared to acceptance of control (unsprayed) eggs. Then laboratory-reared *Trichogramma* were released to test acceptance of laboratory-sprayed egg masses placed in the field. Survival/success in the field was also tested immediately following field application of sulfur by assessing rates of parasitism of sprayed and control egg masses (Figure 4.3).

The results support the suggestion that sulfur is harmful for *T. carverae*. The toxicity of fresh or dried sulfur found in the laboratory trials was confirmed by the lack of survival of adults released in the field trials. The failure of field releases in these spray trials indicates a fairly severe cost to *Trichogramma* released at the time of sulfur application. Sulfur will also affect *Trichogramma* already in the vineyard by reducing rates of emergence from eggs parasitized prior to spray application and decreasing fecundity in those females that do emerge.

If *T. carverae* are to succeed as a replacement for chemical controls, release strategies that work in combination with selective fungicides and miticides need to be developed. Clearly, optimal field performance of released *Trichogramma* requires spraying of sulfur to be undertaken separately from wasp releases. We have undertaken experiments to determine the persistence of the toxic effects of the dried residue of sulfur in the days following spraying: laboratory trials and field releases indicate that the effects of sulfur are sufficiently diminished by day 6 to allow successful *Trichogramma* releases (Thomson *et al.*, 2000).

Conclusions

To address the inherent variability, each pest system must be studied both in the laboratory and in the field before a cost–benefit analysis can be completed. Maximum fecundity on the target host, competence at the required temperature (is acclimation appropriate?), and the optimum rearing host to produce wasps of high fitness and fecundity are key considerations. Different rearing procedures can result in important variation in the quality of the *Trichogramma*. This variation, as well as genetic changes in continuous laboratory cultures, may have significant effects on laboratory-reared parasites to be released in the field. The economics of production may favor producing and releasing greater numbers of less-efficient parasitoids to obtain the same degree of control, although this may not compensate for low quality *Trichogramma* (Dutton *et al.*, 1996). Low numbers of wasps should prove equally effective if large individuals with a high fitness are released (e.g., Bennett and Hoffmann, 1998). The cost–benefit analysis for each system must include the cost of producing the optimum *Trichogramma* for the situation and investigate whether the same result can be achieved more economically by producing more of a less-effective wasp. Field trials are essential to confirm that laboratory tests of suitability are confirmed prior to release. Finally, research is needed to understand methods of storage and dispersal of wasps to ensure that the best product possible is released at the point of use.

In summary, improvement in the success of *Trichogramma* as a biological control agent requires optimization of the materials used, particularly choice of rearing host for provision of a high-quality product. This product must not be affected by inappropriate cold storage or poor conditions during transit and needs to be released at the optimum time to maximize the number of *Trichogramma* at the time of maximum host density. Releases should be appropriately timed to minimize interaction with toxic pesticides, or else relatively harmless pesticides must be chosen. Most importantly, each system must be studied with the optimal species or strain chosen and the potential of acclimation assessed. Obviously not all these parameters can be chosen to fit a specific case, but in a cost–benefit analysis the cost of each choice can be considered in terms of maximizing benefit. A more sophisticated/integrated approach is required in which all aspects of the host–parasitoid interaction as well as the physical conditions such as temperature or presence of pesticides need to be considered.

The study available also illustrates how field trials with sentinel cards in target crops can be used as a quick and inexpensive method of gaining information on the potential for commercial use of *Trichogramma*. Small trials were used to assess detailed aspects of *Trichogramma* performance under field conditions, including dispersal, release rates, spacing of release sites, quality of strains after mass-rearing, and comparisons of strains.

Preliminary field trials have shown that the potential exists for commercial use of *T. carverae* in grapevines, but previous studies over many years clearly illustrate that the test of commercial viability depends on a continuing commitment to strain improvement, quality control, and integration of *Trichogramma* into IPM systems.

Linda Thomson, David Bennett, DeAnn Glenn, and Ary Hoffmann

References

Andow, D.A., Klacan, G.C., Bach, D. and Leahy, T.C. (1995) Limitations of *Trichogramma nubilale* (Hymenoptera: Trichogrammatidae) as an inundative biological control of *Ostrinia nubilalis* (Lepidoptera: Crambidae). *Environ. Entomol.*, **24**, 1352–1357.

Ashley, T.R., Gonzales, D. and Leigh, T.F. (1973) Reduction in effectiveness of laboratory reared *Trichogramma*. *Environ. Entomol.*, **2**, 1069–1073.

Ashley, T.R., Gonzales, D. and Leigh, T.F. (1974) Selection and hybridization of *Trichogramma*. *Environ. Entomol.*, **3**, 43–48.

Bai, B., Cobanoglu, S. and Smith, S.M. (1995) Assessment of *Trichogramma* species for biological control of forest lepidopteran defoliators. *Entomol. Exp. Appl.*, **75**, 135–143.

Bartlett, B.R. (1963) The contact toxicity of some pesticide residues to Hymenopterous parasites and Coccinellid predators. *J. Econ. Entomol.*, **56**, 694–698.

Bennett, D.M. and Hoffmann, A.A. (1998) Effects of size and fluctuating asymmetry on field fitness of the parasitoid *Trichogramma carverae* (Hymenoptera: Trichogrammatidae). *J. Anim. Ecol.*, 67, 580–591.

Bentur, J.S., Kalode, M.B., Rajendran, B. and Patel, V.S. (1994) Field evaluation of the egg parasitoid *Trichogramma japonicum* Ash. (Hym., Trichogrammatidae) against the rice leaf folder, *Cnaphalocrocis medinalis* (Guen.) (Lep., Pyralidae) in India. *J. Appl. Entomol.*, **117**, 257–261.

Bigler, F. (1994) Quality control in *Trichogramma* production. In E. Wajnberg and S.A. Hassan (eds.), *Biological Control with Egg Parasitoids*, CAB International, Wallingford, UK, pp. 93–110.

Bigler, F. and Brunetti, R. (1986) Biological control of *Ostrinia nubilalis* Hubner by *Trichogramma maidis* Pinto et Voeg. on corn for seed production in southern Switzerland. *J. Appl. Entomol.*, **102**, 303–306.

Bigler, F., Bieri, M., Fritschy, A. and Seidel, K. (1988) Variation in locomotion between laboratory strains of *Trichogramma maidis* and its impact on parasitism of eggs of *Ostrinia nubilalis* in the field. *Entomol. Exp. Appl.*, **49**, 283–290.

Bigler, F., Bosshart, S., Frei, G. and Cerutti, F. (1993) Product control of *Trichogramma brassicae* in Switzerland. *Proceedings of the 7th Workshop of the Global IOBC Working Group "Quality Control of Mass Reared Arthropods,"* International Organization of Biological Control, Rimini.

Bjorksten, T.A. and Hoffmann, A.A. (1995) Effects of pre-adult and adult experience on host acceptance in choice and non-choice tests in two strains of *Trichogramma*. *Entomol. Exp. Appl.*, **76**, 49–58.

Bjorksten, T.A. and Hoffmann, A.A. (1998) Separating the effects of experience, size, egg load and genotype on host response in *Trichogramma* (Hymenoptera: Trichogrammatidae). *J. Insect Behav.*, **11**, 129–148.

Boldt, P.E. (1974) Temperature, humidity, and host: effect on rate of search of *Trichogramma evanescens* and *T. minutum* Auctt. (not Riley, 1871). *Ann. Entomol. Soc. Am.*, **67**, 706–708.

Bonhof, M.J., Overholt, W.A., Van Huis, A. and Polaszek, A. (1997) The natural enemies of cereal stemborers in East Africa: a review. *Insect Sci. Applic.*, **17**, 19–35.

Bourchier, R.S. and Smith, S.M. (1996) Influence of environmental conditions and parasitoid quality on field performance of *Trichogramma minutum*. *Entomol. Exp. Appl.*, **80**, 461–468.

Carver, M. (1978) A new subgenus and species of *Trichogramma* Westwood (Hymenoptera: Chalcidoidea) from Australia. *J. Aust. Entomol. Soc.*, **17**, 109–112.

Cerutti, F. and Bigler, F. (1995) Quality assessment of *Trichogramma brassicae* in the laboratory. *Entomol. Exp. Appl.*, **75**, 19–26.

Chang, Y.-F., Tauber, M.J. and Tauber, C.A. (1996) Reproduction and quality of F1 offspring in *Chrysoperla carnea*: differential influence of quiescence, artificially-induced diapause and natural diapause. *J. Insect Physiol.*, **42**, 521–528.

Chihrane, J. and Lauge, G. (1996) Loss of parasitization efficiency of *Trichogramma brassicae* (Hym.: Trichogrammatidae) under high-temperature conditions. *Biol. Cont.*, **7**, 95–99.

Chihrane, J., Lauge, G. and Hawlitzky, N. (1993) Effects of high temperature shocks on the development and biology of *Trichogramma brassicae* (Hymenoptera: Trichogrammatidae). *Entomophaga*, **38**, 185–192.

Consoli, F.I. and Parra, J.R.P. (1995) Effects of constant and alternating temperatures on *Trichogramma galloi* Zucchi (Hymenoptera: Trichogrammatidae) biology. I. Development and thermal requirements. *J. Appl. Entomol.*, **119**, 415–418.

Consoli, F.L., Parra, J.R.P. and Hassan, S.A. (1998) Side-effects of insecticides used in tomato fields on the egg parasitoid *Trichogramma pretiosum* Riley (Hymenoptera: Trichogrammatidae), a natural enemy of *Tuta absoluta* (Meyrick) (Lepidoptera: Gelechiidae). *J. Appl. Entomol.*, **122**, 43–47.

Corrigan, J.E. and Laing, J.E. (1994) Effects of the rearing host species and the host species attacked on performance by *Trichogramma minutum* Riley (Hymenoptera: Trichogrammatidae). *Environ. Entomol.*, **23**, 755–760.

Danthanarayana, W. (1975) The bionomics, distribution and host range of the light brown apple moth, *Ephiphyas postvittana* (Walk.) (Tortricidae). *Aust. J. Zool.*, **23**, 419–437.

Danthanarayana, W. (1976) Flight thresholds and seasonal variation in flight activity of the light brown apple moth, *Epiphyas postvittana* (Walker) (Tortricidae), in Victoria, Australia. *Oecologia*, **23**, 271–282.

De Sa, L.A.N. and Parra, J.R.P. (1994) Natural parasitism of *Spodoptera frugiperda* and *Helicoverpa zea* (Lepidoptera: Noctuidae) eggs in corn by *Trichogramma pretiosum* (Hymenoptera: Trichogrammatidae) in Brazil. *Fla. Entomol.*, **77**, 185–188.

Delpuech, J.M., Froment, M.B., Fouillet, P., Pompanon, F., Janillon, S. and Bouletreau, M. (1998a) Inhibition of sex pheromone communications of *Trichogramma brassicae* (Hymenoptera) by the insecticide chlorpyrifos. *Environ. Toxicol. Chem.*, **17**, 1107–1113.

Delpuech, J.M., Gareau, E., Terrier, O. and Pouillet, P. (1998b) Sublethal effects of the insecticide chlorpyrifos on the sex pheromonal communication of *Trichogramma brassicae*. *Chemosphere*, **36**, 1775–1785.

Delpuech, J.M., Legallet, B., Terrier, O. and Fouillet, P. (1999) Modification of the sex pheromonal communication of *Trichogramma brassicae* by a sublethal dose of deltamethrin. *Chemosphere*, **38**, 729–739.

Dutton, A. and Bigler, F. (1995) Flight activity assessment of the egg parasitoid *Trichogramma brassicae* (Hymenoptera: Trichogrammatidae) in laboratory and field conditions. *Entomophaga*, **40**, 223–233.

Dutton, A., Cerutti, F. and Bigler, F. (1996) Quality and environmental factors affecting *Trichogramma brassicae* efficiency under field conditions. *Entomol. Exp. Appl.*, **81**, 71–79.

Forsse, E., Smith, S.M. and Bourchier, R.S. (1992) Flight initiation in the egg parasitoid *Trichogramma minutum*: effects of ambient temperature, mates, food, and host eggs. *Entomol. Exp. Appl.*, **62**, 147–154.

Franz, J.M., Bogenschutz, H., Hassan, S.A., Huang, P., Naton, E., Suter, H. and Viggiani, G. (1980) Results of a joint pesticide test programme by the working group: pesticides and beneficial arthropods. *Entomophaga*, **25**, 231–236.

Frei, G. and Bigler, F. (1993) Fecundity and host acceptance tests for quality control of *Trichogramma brassicae*. *Proceedings of the 7th Workshop of the Global IOBC Working Group "Quality Control of Mass Reared Arthropods,"* International Organization of Biological Control, Rimini, pp. 81–95.

Glenn, D.C. and Hoffmann, A.A. (1997) Developing a commercially viable system for biological control of light brown apple moth (Lepidoptera: Tortricidae) in grapes using endemic *Trichogramma* (Hymenoptera: Trichogrammatidae). *J. Econ. Entomol.*, **90**, 370–382.

Glenn, D.C., Hercus, M.J. and Hoffmann, A.A. (1997) Characterising *Trichogramma* (Hymenoptera: Trichogrammatidae) species for biocontrol of light brown apple moth (Lepidoptera: Tortricidae) in grapevines in Australia. *Ann. Entomol. Soc. Am.*, **90**, 128–137.

Godfray, H.C.J. (1994) *Parasitoids. Behavioral and Evolutionary Ecology.* Princeton University Press, Princeton, NJ.

Greenberg, S.M., Legaspi, J.C., Nordlund, D.A., Wu, Z.X., Legaspi, B. and Saldana, R. (1998a) Evaluation of *Trichogramma* spp. (Hymenoptera: Trichogrammatidae) against two pyralid stemborers of Texas sugarcane. *J. Entomol. Sci.*, **33**, 158–164.

Greenberg, S.M., Morrison, R.K., Nordlund, D.A. and King, E.G. (1998b) A review of the scientific literature and methods for production of factitious hosts for use in mass rearing of *Trichogramma* spp. (Hymenoptera: Trichogrammatidae) in the former Soviet Union, the United States, Western Europe and China. *J. Entomol. Sci.*, **33**, 15–32.

Greenberg, S.M., Nordund, D.A. and Wu, Z. (1998c) Influence of rearing host on adult size and ovipositional behavior of mass produced female *Trichogramma minutum* Riley and *Trichogramma pretiosum* Riley (Hymenoptera: Trichogrammatidae). *Biol. Cont.*, **11**, 43–48.

Grimm, M. and Lawrence, J.T. (1975) Biological control of the insects on the Ord. I. Production of *Sitotroga cerealella* for mass rearing of *Trichogramma* wasps. *J. Agric. West. Aust.*, **16**, 90–92.

Gunie, G. and Lauge, G. (1997) Effects of high temperatures recorded during diapause completion of *Trichogramma brassicae* prepupae (Hymenoptera: Trichogrammatidae), on the treated generation and its progeny. *Entomophaga*, **42**, 326–329.

Hassan, S.A. (1994) Strategies to select *Trichogramma* species for use in biological control. In E. Wajnberg and S.A. Hassan (eds.), *Biological Control with Egg Parasitoids*, CAB International, Wallingford, UK, pp. 55–72.

Hassan, S.A., Bigler, F., Blaisinger, P., Bogenschutz, H., Boller, E., Brun, J., Chiverton, P., Edwards, P., Huang, P., Inglesfield, C., Naton, E., Oomen, P.A., Overmeer, W.P., Rieckmann, W., Samsoe-Petersen, L., Staubli, A., Tuset, J.J., Viggiani, G. and Vanwetswinkel, G. (1987) Results of the third joint pesticide testing programme carried out by the IOBC/WPRS Working group "Pesticides and Beneficial Organisms." *J. Appl. Entomol.*, **103**, 92–107.

Hassan, S.A., Kohler, E. and Rost, W.M. (1988) Mass production and utilization of *Trichogramma*: 10. Control of the codling moth *Cydia pomonella* and the summer fruit tortrix moth *Adoxophyes orana* (Lepidoptera: Tortricidae). *Entomophaga*, **33**, 413–420.

Hassan, S.A., Bigler, F., Bogenschutz, H., Boller, E., Brun, J., Callis, J.N.M., Coremans-Pelseneer, J., Duso, C., Grove, A., Heimbach, U., Helyer, N., Hokkanen, H., Lewis, G.B., Mansour, F., Moreth, L., Polgar, L., Samsoe-Petersen, L., Suphanor, B., Staubli, A., Stern, G., Vainio, A., Van De Veire, M., Viggiani, G. and Vogt, H. (1994) Results of the sixth joint pesticide testing programme of the IOBC/WPRS-working group "Pesticides and Beneficial Organisms." *Entomophaga*, **39**, 107–119.

Hassan, S.A., Hafes, B., Degrande, P.E. and Herai, K. (1998) The side effects of pesticides on the egg parasitoid *Trichogramma caecoeciae* Marchal (Hymenoptera: Trichogrammatidae), acute dose–response and persistence tests. *J. Appl. Entomol.*, **122**, 569–573.

Hawlitzky, N., Dorville, F.M. and Vaillant, J. (1994) Statistical study of *Trichogramma brassicae* efficiency in relation with characteristics of the European corn borer egg masses. *Res. Popul. Ecol.*, **36**, 79–85.

Hoffmann, A.A. (1995) Acclimation: increasing survival at a cost. *Trends Ecol. Evol.*, **10**, 1–2.

Hoffmann, A.A. and Hewa-Kapuge, S. (2000) Acclimation for heat resistance in *Trichogramma* nr. *brassicae*: can it occur without costs? *Funct. Ecol.*, **14**, 55–60.

Huey, R.B. and Berrigan, D. (1996) Testing evolutionary hypotheses of acclimation. In I.A. Johnston and A.F. Bennett (eds.), *Phenotypic and Evolutionary Adaptation to Temperature*, Cambridge University Press, Cambridge, UK, pp. 205–237.

Huey, R.B., Berrigan, D., Gilchrist, G.W. and Herron, J.C. (1999) Testing the adaptive significance of acclimation; a strong inference approach. *Am. Zool.*, **39**, 323–336.

Hutchinson, L.A. and Bale, J.S. (1994) Effects of sublethal cold stress on the aphid *Rhopalosiphum padi*. *J. Appl. Ecol.*, **31**, 102–108.

Jalali, S.K. and Singh, S.P. (1992) Differential response of four *Trichogramma* species to low temperatures for short term storage. *Entomophaga*, **37**, 159–165.

Kazmer, D.J. and Luck, R.F. (1995) Field tests of the size-fitness hypothesis in the egg parasitoid *Trichogramma pretiosum*. *Ecology*, **76**, 412–425.

Lewis, W.J., van Lenteren, J.C., Phatak, S.C. and Tumlinson, J.H. (1997) A total system approach to sustainable pest management. *Proc. Natl. Acad. Sci., USA*, **94**, 12243–12248.

Li, L.-Y. (1994) Worldwide use of *Trichogramma* for biological control on different crops: a survey. In E. Wajnberg and S.A. Hassan (eds.), *Biological Control with Egg Parasitoids*, CAB International, Wallingford, UK, pp. 37–53.

Madge, D.G. and Stirrat, S.C. (1991) Development of a day degree model to predict generation events for lightbrown apple moth *Epiphyas postvittana* (Walker) (Lepidoptera: Tortricidae) on grapevines in Australia. *Plant Prot. Quart.*, **6**, 39–42.

Maisonhaute, C., Chihrane, J. and Lauge, G. (1999) Induction of thermotolerance in *Trichogramma brassicae* (Hymenoptera: Trichogrammatidae). *Biol. Cont.*, **28**, 116–122.

McDougall, S.J. and Mills, N.J. (1997) The influence of hosts, temperature and food sources on the longevity of *Trichogramma platneri*. *Entomol. Exp. Appl.*, **83**, 195–203.

McLaren, I.W. and Rye, W.J. (1983) The rearing, storage and release of *Trichogramma ivelae* Pang and Chen (Hymenoptera: Trichogrammatidae) for control of *Heliothis punctiger* Wallengren (Lepidoptera: Noctuidae) on tomatoes. *J. Aust. Entomol. Soc.*, **22**, 119–124.

Mertz, B.P., Fleischer, S.J., Calvin, D.D. and Ridgway, R.L. (1995) Field assessment of *Trichogramma brassicae* (Hymenoptera: Trichogrammatidae) and *Bacillus thuringiensis* for control of *Ostrinia nubilalis* (Lepidoptera: Pyralidae) in sweet corn. *J. Econ. Entomol.*, **88**, 1616–1625.

Miura, K. and Kobayashi, M. (1998) Effects of host-egg age on the parasitism by *Trichogramma chilonis* Ishii (Hymenoptera: Trichogrammatidae) an egg parasitoid of the diamondback moth. *Appl. Entomol. Zool.*, **33**, 219–222.

Naranjo, S.E. (1993) Life History of *Trichogrammatoidea bactrae* (Hymenoptera: Trichogrammatidae), an egg parasitoid of pink bollworm (Lepidoptera: Gelechiidae), with emphasis on performance at high temperatures. *Environ. Entomol.*, **22**, 1051–1059.

Newton, P.J. and Odendaal, W.J. (1990) Commercial inundative releases of *Trichogrammatoidea cryptophlebiae* (Hym.: Trichogrammatidae) against *Cryptophlebia leucotreta* (Lep.: Tortricidae) in citrus. *Entomophaga*, **35**, 545–556.

Numata, H. (1993) Induction of adult diapause and of low and high reproductive states in a parasitoid wasp, *Ooencyrtus nezarae*, by photoperiod and temperature. *Entomol. Exp. Appl.*, **66**, 127–134.

Oatman, E.R. and Pinto, J.D. (1987) A taxonomic review of *Trichogramma* (*Trichogrammanza*) Carver (Hymenoptera: Trichogrammatidae), with descriptions of two new species from Australia. *J. Aust. Entomol. Soc.*, **26**, 193–201.

O'Neil, R.J., Giles, K.L., Obrycki, J.J., Mahr, D.L., Legaspi, J.C. and Katovichy, K. (1998) Evaluation of the quality of four commercially available natural enemies. *Biol. Cont.*, **11**, 1–8.

Pak, G.A. and van Heiningen, T.G. (1985) Behavioural variations among strains of *Trichogramma* spp.: adaptability to field-temperature conditions. *Entomol. Exp. Appl.*, **38**, 3–13.

Palmer, A.R. (1994) Fluctuating asymmetry analyses: a primer. In A. Markow (ed.), *Developmental Instability: Its Origins and Evolutionary Implications*. Kluwer Academic Publishers, Dordrecht, pp. 335–364.

Pavlik, J. (1990) The oviposition activity of *Trichogramma* spp. The effect of temperature. In *Trichogramma and Other Egg Parasitoids*, Les Colloques d'INRA, Paris, San Antonio, TX, USA, pp. 85–87.

Pavlik, J. (1993) The size of the female and quality assessment of mass-reared *Trichogramma* spp. *Entomol. Exp. Appl.*, **66**, 171–177.

Ramesh, B. and Baskaran, P. (1996) Developmental response of four species of *Trichogramma* (Hym.: Trichogrammatidae) to heat shocks. *Entomophaga*, **41**, 267–277.

Romeis, J. and Shanower, T.G. (1996) Arthropod natural enemies of *Helicoverpa armigera* (Hubner) (Lepidoptera: Noctuidae) in India. *Biocont. Sci. Technol.*, **6**, 481–508.

Romeis, J., Shanower, T.G. and Jyothirmayi, K.N.S. (1998) Constraints on the use of *Trichogramma* egg parasitoids in biological control programmes in India. *Biocont. Sci. Technol.*, **8**, 289–299.

Schmuck, R., Mager, H., Kunast, C., Bock, K.-D. and Storck-Weyermuller, S. (1996) Variability in the reproductive performance of beneficial insects in standard laboratory toxicity assays — Implications for hazard classification of pesticides. *Ann. Appl. Biol.*, **128**, 437–451.

Scholler, M., Hassan, S.A. and Reichmuth, C. (1996) Efficacy assessment of *Trichogramma evanscens* and *T. embryophagum* (Hymenoptera: Trichogrammatidae) for control of stored products moth pests in bulk wheat. *Entomophaga*, **41**, 125–132.

Scholz, B.C.G., Monsour, C.J. and Zalucki, M.P. (1998) An evaluation of selective *Helicoverpa armigera* control options in sweet corn. *Aust. J. Exp. Agric.*, **38**, 601–607.

Schumacher, P.D., Weber, C., Hagger, C. and Dorn, S. (1997) Heritability of flight distance for *Cydia pomonella. Entomol. Exp. Appl.*, **85**, 169–175.

Scott, M., Berrigan, D. and Hoffmann, A.A. (1997) Costs and benefits of acclimation to elevated temperature in *Trichogramma carverae. Entomol. Exp. Appl.*, **85**, 211–219.

Shipp, J.L. and Wang, K. (1998) Evaluation of commercially produced *Trichogramma* spp. (Hymenoptera: Trichogrammatidae) for control of tomato pinworm, *Keiferia lycopersicella* (Lepidoptera: Gelechiidae), on greenhouse tomatoes. *Can. Entomol.*, **130**, 721–731.

Simser, D. (1994) Parasitism of cranberry fruitworm (*Acrobasis vaccinii;* Lepidoptera: Pyralidae) by endemic or released *Trichogramma pretiosum* (Hymenoptera: Trichogrammatidae). *The Great Lakes Entomol.*, **27**, 189–196.

Smith, S.M. (1994) Methods and timing of releases of *Trichogramma* to control lepidopterous pests. In E. Wajnberg and S.A. Hassan (eds.), *Biological Control with Egg Parasitoids*. CAB International, Wallingford, UK, pp. 113–144.

Smith, S.M. (1996) Biological control with *Trichogramma*: advances, successes and potential of their use. *Annu. Rev. Entomol.*, **41**, 375–406.

Smith, S.M., Hubbes, M. and Carrow, J.R. (1987) Ground releases of *Trichogramma minutum* Riley (Hymenoptera: Trichogrammatidae) against the spruce budworm (Lepidoptera: Tortricidae). *Can. Entomol.*, **119**, 251–263.

Sorati, M., Newman, M. and Hoffmann, A.A. (1996) Inbreeding and incompatibility in *Trichogramma* nr. *brassicae*: evidence and implications for quality control. *Entomol. Exp. Appl.*, **78**, 283–290.

Stinner, R.E. (1977) Efficacy of inundative releases. *Annu. Rev. Entomol.*, **22**, 515–531.

Suckling, D.M., Chapman, R.B. and Penman, D.R. (1984) Insecticide resistance in the light brown apple moth, *Epiphyas postvittana* (Walker) (Lepidoptera: Tortricidae): larval response to Azinphosmethyl. *J. Econ. Entomol.*, **77**, 579–582.

Sundaram, M., Dhandapani, K.N., Swamiappan, M., Babu, P.C.S. and Jayaraj, S. (1994) A study on the management of some pests of ground nuts (*Arachis hypogaea* L.) with biocontrol agents. *J. Biol. Cont.*, **8**, 1–4.

Thomson, L.J., Glenn, D.C. and Hoffmann, A.A. (2000) Effects of sulphur on *Trichogramma* egg parasitoids in vineyards: measuring toxic effects and establishing release windows. *Aust. J. Exp. Agr.*, **40**, 1165–1171.

Thomson, L.J., Robinson, M. and Hoffmann, A.A. (2001) Field and laboratory evidence for acclimation without costs in an egg parasitoid. *Funct. Ecol.*, **15**, 217–221.

Twine, P.H. and Lloyd, R.J. (1982) Observations on the effect of regular releases of *Trichogramma* spp. in controlling *Heliothis* spp. and other insects in cotton. *Queensland J. Agric. Anim. Sci.*, **39**, 159–167.

van Bergelijk, K.E., Bigler, F. and Kaasboek, N.K. (1989) Changes in host acceptance and host suitability as an effect of rearing *Trichogramma maidis* on a factitious host. *Entomol. Exp. Appl.*, **52**, 229–238.

Vasquez, L.A., Shelton, A.M., Hoffmann, M.P. and Roush, R.T. (1997) Laboratory evaluation of commercial Trichogrammatid products for potential use against *Plutella xylostella* (L.) (Lepidoptera: Plutellidae). *Biol. Contr.*, **9**, 143–148.

Linda Thomson, David Bennett, DeAnn Glenn, and Ary Hoffmann

Visser, M.E. (1994) The importance of being large: the relationship between size and fitness in females of the parasitoid *Aphaereta minuta* (Hymenoptera: Braconidae). *J. Anim. Ecol.*, **63**, 963–978.

Voegele, J., Pizzol, J. and Babi, A. (1988) The overwintering of some Trichogramma species. In *Trichogramma and Other Egg Parasitoids*. Les Colloques dÍNRA, Paris, San Antonio, TX, USA, pp. 275–282.

Wajnberg, E. and Collazo, A. (1998) Genetic variability in the area searched by a parasitic wasp: analysis from automatic video tracking of the walking path. *J. Insect Physiol.*, **44**, 437–444.

Wang, Z., Ferro, D.N. and Hosmer, D.W. (1997) Importance of plant size, distribution of egg masses, and weather conditions on egg parasitism of the European corn borer, *Ostrinia nubilalis* by *Trichogramma ostriniae* in sweet corn. *Entomol. Exp. Appl.*, **83**, 337–345.

West, S.A., Flanagan, K.E. and Godfray, H.C.J. (1996) The relationship between parasitoid size and fitness in the field, a study of *Achrysocharoides zwoelferi* (Hymenoptera: Eulophidae). *J. Anim. Ecol.*, **65**, 631–639.

Yu, D.S. and Byers, J.R. (1994) Inundative release of *Trichogramma brassicae* Bezdenko (Hymenoptera: Trichogrammatidae) for control of the European corn borer in sweet corn. *Can. Entomol.*, **126**, 291–301.

Zaslavski, V.A. and Umarova, T.Y. (1990) Environmental and endogenous control of diapause in *Trichogramma* species. *Entomophaga*, **35**, 23–29.

TRITROPHIC INTERACTIONS: THE INDUCIBLE DEFENSES OF PLANTS

5

Dale M. Norris[1] and Ingrid Markovic[2]
[1]University of Wisconsin, Madison, USA, [2]Laboratory of Cell Biology, Division of Monoclonal Antibodies, Center for Biologics Evaluation and Research, Food and Drug Administration, Bethesda, USA

Introduction

The inducible defenses of plants have a genomic-chemical basis, but their ultimate expression in the phenotype may be anatomical, morphological, or chemical (Norris and Kogan, 1980). The extent of their expression in the phenotype is almost if not always influenced by the environment. The environment for this contribution is the one of tritrophic interactions, i.e., emphasizing the plant, the herbivore, and the parasitoid or predator; and a chief interest is how these interactions influence (i.e., alter) the involved plant's abilities to cope (i.e., survive and propagate). Thus, the emphasis in this chapter is on tritrophic interactions, but one example of polytrophic living systems is discussed. Because there are some rather similar topics on tritrophic interactions in this volume, some overlap among chapters may occur. In the present chapter, interpretations of the following subjects are especially offered:

- Dual (intrinsic and extrinsic) alterable plant chemical defenses
- Distinct plant roles in the evolution of kairomones for "generalist" versus "specialist" parasitoids of herbivores
- A common dependence of intrinsic and extrinsic plant chemical defenses
- Multiple communicative roles of individual compounds in a mixture of messenger volatiles

- Polytrophic symbiosis allowing the insect partner to deal with diverse inducible host-plant chemistries at monophagous metabolic costs
- A sulfhydryl (—SH, thiol)/disulfide (—S—S—)-dependent, redox-based communication system involved in inducible plant defenses
- Inducible plant defenses: another communicative realm in biology involving the "environmental energy exchange code"

Dual (intrinsic and extrinsic) alterable plant chemical defenses

The discovery that differing environmental (i.e., extracellular energy-state) conditions could increase, or decrease, the level of a plant's intrinsic chemical defense against such factors (i.e., stresses such as pests and pathogens) was a major contribution to secondary plant physiology and biochemistry. Akazawa *et al.* (1960) showed that insect feeding in plant tissues could elicit defensive metabolism in the plant. Cruickshank and Perrin (1965) studied several factors as elicitors of phytoalexins against pathogenic microorganisms in plants. Kogan and Paxton (1983) reviewed early findings on natural inducers of plant resistance to insects, and Ryan (1983) discussed insect-induced chemical signals regulating plant defensive responses. Sequeira (1983) analyzed mechanisms of induced microbial resistance in plants. Norris and associates (e.g., Chiang *et al.*, 1987; Neupane and Norris, 1990, 1991a,b, 1992; Ramachandran and Norris, 1991; Ramachandran *et al.*, 1991; Haanstad and Norris, 1992; Liu *et al.*, 1988, 1989, 1992; Norris and Liu, 1992; Burden and Norris, 1994; Markovic *et al.*, 1993, 1996a,b, 1997; Norris, 1994a,b) have especially contributed to the following aspects of alterable plant chemical defenses:

- A given quantity of a specific elicitor (i.e., stress) may increase, or decrease, inducible defense chemistry depending on plant species, plant variety or hybrid, season, method of application, time after elicitation, and numerous other environmental factors (i.e., conditions).
- There may be, at least, three orders of alterable chemical defenses in a plant: first, repellent volatiles; second, less volatile feeding deterrents; and third, inhibitors of herbivore general physiology and development.
- Alterable plant defense usually involves changes in the concentration of component chemicals in the plant's genetically determined allelochemical arsenal.
- The host plant usually plays distinct, but essential, roles in the production of kairomones for "generalist" versus "specialist" parasitoids.
- Alterable defense involves a sulfhydryl (—SH, thiol)/disulfide (—S—S—) redox-dependent energy-transduction mechanism (Figure 5.1) in the plant regardless of the specific stress factor involved in its environment.

Dale M. Norris and Ingrid Markovic

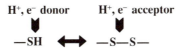

H⁺, e⁻ donor   H⁺, e⁻ acceptor

—SH ⟷ —S—S—

Figure 5.1 The sulfur-based redox couple, sulfhydryl (thiol, —SH) / disulfide (—S—S—), in which the addition of a reducing agent (hydrogen, electrons/protons) favors the formation of thiols, whereas the addition of an oxidizing agent (e.g., quinone) favors the reverse, i.e., formation of disulfides. This redox couple in proteins is considered the electrochemical functional interface by which energy is exchanged as information between living entities and their environments.

Intrinsic alterable plant chemical defenses

Intrinsic chemical plant defenses are defined here as those that act directly on the insect (i.e., the herbivore). Such defenses are attributable to a continuum of phytochemistry (Markovic *et al.*, 1996a,b, 1997) ranging from highly volatile repellents, through less volatile feeding deterrents, to potent inhibitors of survival, development, molting, pupation, and weight gain. Such a continuum of intrinsic defense may thus attack the herbivore at multiple target sites, producing negative insect orientation, reduced feeding, increased developmental time, disrupted molting and pupation, and increased mortality. A plant can use chemicals to yield, at least, three orders of intrinsic defense:

1. Volatiles, which elicit negative insect orientation with regard to the source, and may thus negatively affect the insect before physical contact with the plant occurs
2. Less volatile compounds, which prevent, or reduce, feeding if the insect contacts the plant
3. Chemicals that variously disrupt the herbivore's nutritional and developmental physiology

It is significant that as the insect is exposed sequentially to the first, second, and third orders of intrinsic defense, it incurs increasing costs because the adverse effects of the encountered phytochemistry become less reversible.

It is clear that distinct genes may control these different orders of intrinsic defense. Within soybean, *Glycine max* (Linnaeus) Merrill, the volatiles of the "Davis" commercial variety proved attractive to the adult Mexican bean beetle, *Epilachna varivestis* Mulsant, but larvae that hatched from eggs laid by this beetle on "Davis" all died when presented with only foliage from this variety as food (Burden and Norris, 1994). Consumption of the foliage of "Davis" by attracted adult female *E. varivestis* also caused a premature termination in their egg laying (i.e., aborted reproduction). In sharp contrast, the volatiles of the highly insect-resistant *G. max* breeding introduction PI 227687 strongly repelled the adult *E. varivestis* (Burden and Norris, 1994). Thus, agronomic breeding of *G. max* to yield the commercial variety "Davis" altered the genomic control of volatile chemical metabolism in this soybean such that it became a "death trap" for the Mexican bean beetle. That is, the adults are strongly attracted to "Davis" but hatching larvae all die if they eat the tissues of this plant. The involved breeding of soybean thus removed the first order of plant chemical defense, repellents, but the second and third orders of such defense to this insect remain genetically intact in "Davis." With a preferred leguminous host plant, "Henderson" lima bean, its volatiles proved orientationally neutral to adult Mexican bean beetles (Liu *et al.*, 1989); and, of course, larvae completed growth and development, and adults reproduced (Burden and Norris, 1994). Thus, a preferred host here involved a plant lacking significant levels of all three orders of intrinsic chemical defense. In some situations a host plant (e.g., "Davis" soybean) does produce kairomones for a herbivore, e.g., *Trichoplusia ni* (Hubner) (Liu *et al.*, 1988).

Extrinsic alterable plant chemical defenses

Extrinsic chemical plant defenses are defined here as those that act directly as kairomones on a parasitoid or predator to attract it to the host plant of a potential herbivore host or prey, or more specifically to the potential herbivore host or prey on the plant. Early studies of foraging activities of several entomophagous insects only showed that they were chemically influenced behaviorally by volatiles associated with the general habitats of their hosts (Vinson, 1976, 1981; Weseloh, 1981; van Alphen and Vet, 1986). Kairomones for some entomophagous insects have since been associated with plants bearing a feeding host (Drost *et al.*, 1988; Dicke *et al.*, 1990; Turlings *et al.*, 1990, 1991). The specific biological source of such kairomones associated with plant–host complexes remained poorly defined for several years. Candidate specific sources were listed as the plant, the host insect, or host by-products such as feces, silk, or honeydew (Vinson, 1976; Kennedy, 1984; Mudd *et al.*, 1984; Takabayashi and Takahashi, 1989). Turlings *et al.* (1991), however, conclude further that the plant is the vital source of informational volatiles for the generalist parasitoid *Cotesia marginiventris* (Cresson). Orientation of generalist parasitoids, or predators, to cues associated with the plant fed on by herbivores, rather than relying solely on cues associated with hosts/prey, may enable natural enemies to attack effectively a wider set of species within a given feeding niche (Barbosa, 1998). In contrast, volatiles from host feces have proven more important as kairomones for the specialist parasitoids *Microplitis croceipes* (Cresson) (Elzen *et al.*, 1987; Eller *et al.*, 1988) and *M. demolitor* (Wilkinson) (Ramachandran *et al.*, 1991).

Distinct plant roles in the evolution of kairomones for "generalist" versus "specialist" parasitoids

The combined early studies of kairomones for entomophagous insects, as discussed partially above, left the possibility that there is a dichotomy regarding the vital source of kairomones for such generalist versus specialist parasitoids, i.e., the plant for generalists versus the host insect for the specialist entomophage. Basic logic might lead us to conclude that these two distinct sources of kairomones for generalists versus specialists, respectively, meet the essential communication requirements of each class of parasitoid.

However, findings from subsequent more detailed experiments (Ramachandran *et al.*, 1991) strongly support the attractiveness of kairomones from both the plant tissues and the herbivore host's frass (i.e., 3-octanone) and from just the herbivore host's frass (i.e., guaiacol). This has been specifically seen in the location by the specialist parasitoid *M. demolitor* and arrestance on the host larva of the soybean looper, *Pseudoplusia includens* (Walker). In these studies, *M. demolitor* was especially arrested, as well as attracted, by guaiacol. A comparison of electroantennogram (EAG) responses of the host herbivore, *P. includens*, and the specialist parasitoid, *M. demolitor*, to aliphatic compounds of varying carbon-chain lengths (Ramachandran and Norris, 1991) showed that the herbivore's antennal response peaked at six carbons, whereas the parasitoid's peaked at seven and eight carbons. It was suggested that the herbivore's EAG-response profile was adaptive, as six-carbon aliphatic compounds are abundant in the volatiles of its host plant, *G. max* (Liu *et al.*, 1989). Because seven- and eight-carbon volatiles are less common in such plants, Ramachandran and Norris (1991) speculated that the parasitoid's antennal receptors are tuned preferentially to perceive volatiles from other sources in addition to plants. The frass of its herbivore host, *P. includens*, proved to be a source of seven- and eight-carbon volatiles for this parasitoid (Ramachandran *et al.*, 1991). Thus, the specialist parasitoid, *M. demolitor*, was attracted by the volatile, 3-octanone, found in the host plant, *G. max*, of its herbivore host, *P. includens*; however, it was attracted, and markedly more arrested, by the volatile, guaiacol, which was found only in its host's frass. In addition, 3-octanone was found in concentrations in host frass that were ca. 10 times those found in the studied host plants of the herbivore. This finding indicates that this eight-carbon plant-borne volatile to which *M. demolitor's* antennal receptors are especially sensitive was at least highly accumulated and concentrated in the herbivore's frass, making the latter more attractive to the parasitoid. Although generalist parasitoids reportedly respond to volatiles from the host plant of their host herbivore (Turlings *et al.*, 1990, 1991), the complete origins of kairomones for generalist entomophages remain somewhat ill defined; however, they are apparently distinct from the clarified situation for at least one specialist entomophage, *M. demolitor*. In this latter case the specialist parasitoid *M. demolitor* responds not only to compounds from the plant but especially to a kairomone detected only in the host herbivore's frass.

A common dependence of intrinsic and extrinsic plant chemical defenses

There has long been a significant theoretical argument that plant resistance to herbivores (PRH) and biological control of herbivores (BCH) are compatible components in an integrated pest management (IPM) program (Horber, 1972; Adkisson and Dyck, 1980). The findings of Ramachandran *et al.* (1991), however, not only showed that chemically based PRH and BCH for a herbivore, *P. includens*, are compatible, but that both are dependent upon the plant's chemistry. 3-Octanone, a kairomone to both the herbivore host, *P. includens* (Liu *et al.*, 1989), and its specialist parasitoid, *M. demolitor* (Ramachandran *et al.*, 1991), was identified from volatiles collected from intact foliage of such host plants of *P. includens* as "Davis" and PI 227687 *G. max* and lima beans (Liu *et al.*, 1989; Ramachandran *et al.*, 1991). The concentration of 3-octanone in *P. includens* frass produced on such host plants, however, was ca. 10 times greater than in the intact plant tissues; thus, the herbivore's metabolism of plant tissues produces marked 3-octanone concentration that appears to be adaptive to *M. demolitor*. The essential role of the plant as the source of 3-octanone, or a precursor from which the herbivore's metabolism could make this kairomone, was shown clearly by results in which the herbivore produced 3-octanone in its frass when fed host plant tissues but failed to do so when fed only an artificial diet (Ramachandran *et al.*, 1991). Earlier studies (e.g., Sauls *et al.*, 1979; Nordlund and Sauls, 1981; Elzen *et al.*, 1984; Nordlund and Lewis, 1985) showing that several parasitoids failed to respond to volatiles produced in the frass of host herbivores fed on artificial diet surely suggested that the host plant might play a critical role in the herbivore's ability to produce kairomones for a parasitoid.

Guaiacol also was only detected in *P. includens* frass when this herbivore was fed on host plants, but was never detected in volatiles from the involved intact host plants (Ramachandran *et al.*, 1991). The host herbivore thus requires the plant tissues as a source of chemical precursor from which it then produces in its frass the potent kairomone, guaiacol, of the specialist parasitoid *M. demolitor*. Axelrod and Tomchick (1958) showed that guaiacol is readily produced from the common phytochemical catechol, by enzymatic *O*-methylation in animal systems.

A positive correlation between the levels of intrinsic and extrinsic plant chemical defenses to insect herbivores was determined for several leguminous hosts (Ramachandran *et al.*, 1991; Liu *et al.*, 1988, 1989, 1992). The latter authors established that the order of intrinsic defense among the plants was PI 227687 soybean > "Davis" soybean > lima bean. The former workers found the same order for the amount of the parasitoid kairomone, quaiacol, produced in the frass of the herbivore when it fed on the respective plants. Liu *et al.* (1992) further showed that the relative total chemical resistance among several legumes is so strongly regulated genetically that the order of relative defense can be very effectively predicted for the fully developed plant by relatively simple chemical analyses of hypocotyls of seedlings that have been germinated for only a few (e.g., 4–6) days. Use of this new rapid hypocotyl-based chemical analysis allowed an improved classification of the relative total chemical defense in six numerically coded soybean genotypes previously ranked for such defense on the basis of the career-long accumulated knowledge of the renowned soybean breeder, and co-author, Dr. E.E. Hartwig (Liu *et al.*, 1992).

Dale M. Norris and Ingrid Markovic

Multiple communicative roles of individual compounds in a mixture of messenger volatiles

The, at least, dual (i.e., allomonic versus kairomonic) roles of a chemical compound as a messenger in chemical ecology have been emphasized for several decades (Norris and Kogan, 1980). This now thoroughly documented situation in chemical ecology and communication brings to ecology an elegantly simple chemically based mechanism for ordering the distributions, and abundances, of species, subspecies, biotypes, varieties, etc., in time and space. Two further important facts are the following.

- A given compound is commonly just one of several components in a natural chemical communication cue (e.g., Liu *et al.*, 1988, 1989; Markovic *et al.*, 1993) (Figure 5.2).
- A given compound commonly has a variety of allomonic and kairomonic messenger functions across the plant, animal, and microbial realms (Rodriquez and Levin, 1976; Norris and Liu, 1992; Raina *et al.*, 1992).

The long-held common opinion of the "rather sacred" uniqueness of pheromones among chemical messengers thus lacks basis in fact (e.g., Raina *et al.*, 1992). Being a pheromone is just one of several messenger roles a compound may serve in chemical ecology.

Further, elaboration of some of the known messenger roles served by some of the volatiles that proved to be major cues in the chemical interactions among several leguminous host plants, the insect herbivore *P. includens*, and the specialist parasitoid *M. demolitor* (Ramachandran *et al.*, 1991) will serve as a satisfactory example of the multiple communicative activities that individual chemicals have in biology and ecology. 3-Octanone, isolated both from the leguminous host plants and the frass of the host herbivore when feeding on such plants, was kairomonic to *M. demolitor*. This chemical has also been reported as a major constituent of the alarm pheromones of several species of ants (Crewe *et al.*, 1972; Crewe and Blum, 1972; Duffield and Blum, 1975). 3-Octanone also is a component of the volatiles released by the highly insect-resistant soybean PI 277687, which proved highly repellent to both the cabbage looper, *T. ni*, and the Mexican bean beetle, *E. varivestis* (Liu *et al.*, 1989). Guaiacol, the volatile that was detected only in *P. includens* frass, and then only when the host herbivore was feeding on a host plant, proved both attractive and especially arrestive to the specialist parasitoid *M. demolitor*. This compound had previously been identified as the male-specific chemical in the abdominal glands of the leaf-footed bug, *Veneza phyllopus* (Linnaeus) (Aldrich *et al.*, 1976). It also was suggested as one of the long-distance male-produced sex pheromones for this bug species. Duffey and Blum (1977) isolated guaiacol from cyanogenic glands of three species of millipedes, and reported that this chemical functioned as a defense compound for the millipedes. Although guaiacol was never detected from the leguminous host plants studied by

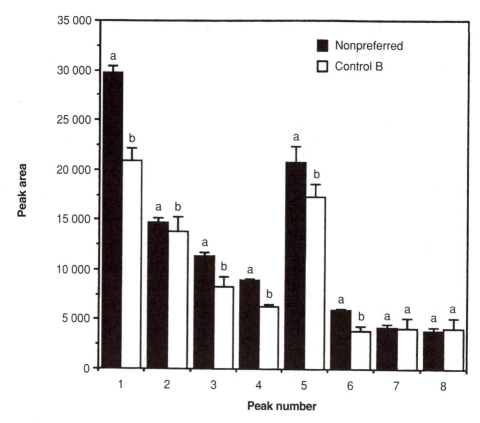

Figure 5.2 Comparative mean areas of the indicated peaks in the standardized high-performance liquid chromatography (HPLC) analysis of the nonhydrolyzed fraction of the ethyl acetate extractables from ash foliage. Data are for nonpreferred foliage (elicited with 25 IU of vitamin E per ml) and control B foliage (elicited only with solvent). Leaves were collected 8 days after elicitation. Areas of Pairs of Peaks with different letters (a,b) are significantly different at $p < 0.05$. (Markovic *et al.*, 1993.)

either Liu *et al.* (1989) or Ramachandran *et al.* (1991), it was reported to occur in numerous plant species (Gibbs, 1974); however, the author could not state whether it occurred in its free form in intact plants.

Another major component volatile from *P. includens* frass produced on host plants is linalool, which was repellent to *M. demolitor* at a high concentration. A perceiving organism (i.e., *M. demolitor*) was thus dealing with a mixture of volatile messengers from a given natural source (*P. includens* frass produced on given leguminous host plants). Its behavioral response was based on a net (allomonic versus kairomonic) messenger input, which in the studied cases of *P. includens* frass was net kairomonic. Linalool was also reported as an important kairomone for the predatory mites *Amblyseius potentillae* (Garman) and *Phytoseiulus persimilis* Athias-Henriot (Dicke *et al.*, 1990). In this study lima beans infested with the herbivorous mite *Tetranychus urticae* Koch produced large quantities of linalool. The authors also suggested that linalool is a component of the *T. urticae* dispersing pheromone. Linalool also has been identified in the metathoracic gland secretions of the cotton-stainer bug, *Dysdercus superstitiosus* (Fabricius) (Daroogheh and Olagbemiro, 1982). In this case it was hypothesized to be a defense compound.

Polytrophic symbiosis allowing the insect partner to deal with diverse host-plant chemistries at monophagous metabolic prices

Partially because of the diversity of plant defensive chemistries, extreme polyphagy is not common among insect herbivores. Some insects, however, exhibit little, if any, discrimination regarding host plant species through their elaborate mutualism with microbial ecto- and endosymbiotes. Some species in the scolytid genus *Xyleborus*, and especially *X. ferrugineus* (Fabricius), are prime examples of such behavior regarding plant substrate (Samuelson, 1981). Wood (1982), a world authority on scolytid beetles, concluded that among species in this family their breadth of acceptable botanical substrate correlates strongly with their evolved nutritional dependence on fungal ectosymbiotes (i.e., mycetophagy) versus phloemphagy. As an excellent example, *X. ferrugineus* adult females cultivate and transport in special ectodermal organs, termed mycangia, a defined (Baker and Norris, 1968) complex of ectosymbiotic fungi that the female inoculates on the walls of the brood chamber she excavates in the sapwood (inner tissues) of whatever plant substrate is available. She then actively cultivates this complex of ectosymbiotes on the chamber walls as the sole food for her and her brood of progeny. In excavating the brood chamber, the adult female does not ingest a significant quantity of the plant substrate; she merely bites out the prescribed cavity and literally kicks the substrate pieces out of the chamber behind her. She is not able to lay eggs until she has the symbiotic microbes growing on the chamber walls and consumes some as food. She must have the microbial growth available as food when her progeny larvae hatch.

Thus, regardless of what plant species the adult female *X. ferrugineus* uses as substrate, her progeny larvae are monophagous in consuming the same ectosymbiotic microbial complex involving *Fusarium solani*, *Graphium* sp., and *Cephalosporium* sp. (Baker and Norris, 1968). These fungi in the Fungi Imperfecti are commonly vascular pathogens of a wide range of plant species, and can grow on (in) a very broad variety of botanical species and still provide the insects with a quite uniform food substrate. Norris (1992) thus described this obligate symbiotic interrelationship as providing this insect with "polyphagous privileges" at "monophagous metabolic prices"; and with more than adequate attributes to qualify as an ideal extreme biofacies (Hamilton, 1967). Such an organism is one that achieves "maximal" conservation of its available energy for the perpetuation of its genes (Norris, 1992).

In extensive chemosensitive studies, *X. ferrugineus* adult females were non-responsive to an extensive variety of widely allomonic phytochemicals. This lack of response included the potent allelochemical juglone (5-hydroxy-1,4-naphthoquinone) (Gilbert *et al.*, 1967). However, the adult females are highly attracted to (i.e., "alcoholic" toward) the major fermentation product, ethanol (Norris and Baker, 1969). Their attack of field-exposed, uniform-sized (10 cm diameter × 50 cm long) log sections from a host tree, *Theobroma cacao* Linnaeus, began 9 days after the sections were cut from apparently healthy trees, and attack peaked at 18 days. The initiation of beetle attack at 9 days corresponded positively with an obvious degree of fermentation and deterioration of the plant tissues that could be readily detected by the human sense of smell. *X. ferrugineus* occurs worldwide in tropical environments, and is now known to attack especially anaerobically stressed (i.e., fermenting) individuals of very many plant species.

A sulfhydryl (−SH) / disulfide (−S−S−)-dependent, redox-based communication system involved in inducible plant defenses

Scientific evidence supporting a sulfhydryl (−SH, thiol) / disulfide (−S−S−)-dependent, redox-based exchange of energy in cell–cell, virus–cell, and ligand–cell communication is biologically very significant and deserves further study.

First elucidation of energy exchange for receptor function in chemosensitive neurons

The pertinent scientific evidence had its beginnings in the early findings about how chemical messengers (e.g., kairomones, allomones, pheromones) interact mechanistically with receptors of insect primary chemosensitive neurons to exchange energy or chemical groups to predict whole-animal behavior such as chemotaxis, feeding inhibition, mating, etc. (Norris, 1969, 1971, 1977; Norris *et al.*, 1970a,b, 1971; Ferkovich and Norris, 1972; Rozental and Norris, 1973, 1975; Norris and Chu, 1974; Singer *et al.*, 1975a,b). This body of research showed that several 1,4-naphthoquinone allelochemicals caused the same order of relative repellency and deterrency to two very different insects (i.e., *Scolytus multistriatus* (Marsham) (Coleoptera: Scolytidae) and *Periplaneta americana* (Linnaeus) (Orthoptera: Blattidae)) based on several behavioral assays. This order was 5-hydroxy-1,4-naphthoquinone > 1,4-naphthoquinone > 2-hydroxy-1,4-naphthoquinone > 2-methyl-1,4-naphthoquinone. In addition:

● The four allelochemicals showed the same order of relative affinity for receptor sites on the thiol/disulfide-dependent energy-transducing protein purified from dendritic nerve membrane of primary chemosensitive neurons in the insect antennae.

- They showed the same order of relative inhibition, in millivolts, of standardized EAGs elicited by the excitant amyl acetate.
- They showed the same order of relative induced $U_{1/2}$ shift in electromotive force (EMF), in millivolts, in a standardized preparation of $100\,\mu g$ of the receptor (energy-transducing) thiol/disulfide protein as measured classically by dropping mercury electrode (DME) polarography.

Three possible pairings of the chemoreception parameters (1)–(3) yielded a linear-regression relationship with $r > 0.95$:

1. The number of moles of messenger naphthoquinone required in a standardized insect behavioral assay to yield a >99 percent change in that behavior.
2. The maximal *in vitro* polarographic $U_{1/2}$ shift, in millivolts, elicited in the involved receptor and energy-transducer protein by saturation with the given messenger.
3. The maximal percent inhibition, in millivolts, of a standardized excitant-stimulated EAG by saturation of the antenna with the given naphthoquinone.

Subsequent simultaneous mathematical solution of the three resultant linear relationships further showed that the determined correlation ($r^2 = 0.95$) between (2) and (3) is so high (i.e., so strong) that only one (either (2) or (3)), as expressible in millivolts, need be considered in a meaningful mathematical description of the thiol/disulfide-dependent redox-electrochemical transduction of the involved abiotic messenger energy from millivolts into quantitated insect behavioral change (i.e., living change).

Numerous independent studies subsequently confirmed the sulfhydryl/disulfide dependence of such conversion of a messenger-borne energy state into chemosensitive neuronal information and organismal behavioral change (e.g., Getchell and Gesteland, 1972; Shimada *et al.*, 1972, 1974; Villet, 1974; Frazier and Heitz, 1975; Mooser, 1976; Ma, 1977, 1981; Vande Berg, 1981).

Supportive evidence involving inducible plant defenses

Classical sulfhydryl reagents (e.g., *p*-chloromercuribenzoic acid (*p*CMB), *p*-chloromercuriphenylsulfonic acid (*p*-CMBS), iodoacetic acid (IAA)), disulfide-reducing agents (e.g., reduced glutathione, dithiothreitol), thiol-oxidizing agents (e.g., 1,4-naphthoquinones), and natural antioxidants (e.g., L-ascorbic acid and α-tocopherol) have been used to alter predictably (i.e., increase or decrease) plant chemical defenses against environmental stresses (e.g., herbivores and pathogens). These findings demonstrate that the qualitative and quantitative exchange of energy or chemical groups as information in situations of alterable plant defenses depends mechanistically on sulfhydryl/disulfide redox-based electrochemistry, which can be readily expressed in millivolts (e.g., Stossel, 1984; Neupane and Norris, 1990, 1991a,b, 1992; Haanstad and Norris, 1992; Liu *et al.*, 1992, 1993; Norris and Liu, 1992; Markovic *et al.*, 1993; Burden and Norris, 1994; Norris 1994a,b).

Inducible plant defenses: another communicative realm in biology involving the "environmental energy exchange code"

The sulfhydryl/disulfide-dependent redox electrochemistry essential to receptor function and energy-transduction mechanisms has become a major area of research throughout biology (van der Vliet and Bast, 1992; Markovic *et al.*, 1998). The early-demonstrated involvement of such electrochemistry in primary-cell neuronal chemoreception and inducible (alterable) plant defenses is now regarded as a mere part of the much larger essential communicative functions of such sulfur-dependent electrochemistry in living systems. The enormity of demonstrated involvement of such sulfur electrochemistry in information generation, transfer, and use in maintaining order in and among cells, and between cells and their abiotic environments, led Norris (1986, 1988, 1994a,b) to theorize that sulfur electrochemistry is a universal messenger in chemical communications among living organisms (animals, humans, plants, microbes, viruses) and their abiotic environments.

An informational code that allows the special use of energy for establishing order must be based on (a) a universal medium (i.e., a messenger) and (b) a means of using that universal messenger to convey both qualitative and quantitative messages (Harold, 1986). The most recent (Norris, 1994a,b) theoretical version of such a code has been termed the "Environmental Energy Exchange Code." In this proposed EEE Code, sulfur electrochemistry is the universal medium (messenger); and the thiol ($-SH$, sulfhydryl), its derivatives, and its redox couple, disulfide ($-S-S-$), are the chemical means of using sulfur electrochemistry to convey both qualitative and quantitative messages. Wald (1969) concluded that the required traits for such chemical messengers single out phosphorus and sulfur from all other atoms in the "Periodic System." The two basic atomic characteristics that uniquely qualify phosphorus and sulfur as agents of the required chemical-group and energy transfers are (a) the possession of d orbitals in addition to s and p orbitals, which increases their capacity to form linkages with a variety of energy potentials; and (b) an intrinsic instability of such linkages, which facilitates the exchange of chemical groups and energy (i.e., informational units) (Wald, 1969).

Such roles of phosphorus have long been well-established experimentally, but they are only more recently being widely recognized for sulfur.

However, sulfur now seems to be the most highly qualified element regarding abilities to exchange chemical groups and energy as information among living forms (including viruses), and their abiotic environments (Wald, 1969; Norris *et al.*, 1970a,b, 1971; Norris, 1969, 1971, 1979, 1981, 1985, 1986, 1994a,b; Rozental and Norris, 1973, 1975; Singer *et al.*, 1975a,b; Neupane and Norris, 1992; Norris and Liu, 1992; Van der Vliet and Bast, 1992; Markovic *et al.*, 1998).

- Regarding such roles, sulfur, first as thiol ($-SH$), brings to living systems a form of organic sulfur that possesses atomic traits otherwise limited to organic oxygen, and which are so essential to organisms. However, thiol lacks most, if not all, of those traits (or degrees thereof) possessed by oxygen that make oxygen or its free-radical derivatives especially harmful, even deadly, to living organisms, unless the oxygen is handled in special ways (e.g., bound to hemoglobin in vertebrate blood (Wald, 1969; Harold, 1986).

- Sulfur, as well as phosphorus, forms relatively long (i.e., loose) bonds with other atoms; such bonds hold the attached atom less tightly and thus it may be more easily transferred or exchanged, as informational energy in living systems (Wald, 1969; Harold, 1986).

- The trait that ultimately sets sulfur apart from phosphorus as a vehicle of chemical-group and energy transfers as information, is the sulfur's ability to form the thiol ($-$SH). This property uniquely enables sulfur to readily accept, donate, or variously share the basic energy unit of living systems, hydrogen (Wald, 1969; Szent-Gyorgyi, 1973).

Thus, through sulfhydryl ($-$SH)/disulfide ($-$S$-$S$-$) redox chemistry, sulfur uniquely brings to living cells, and their chemical communications, a "give and take" electrochemical mechanism for moving hydrogens (electrons/protons), the basic energy unit of living systems, as information among plants, animals including humans, and microbes including viruses, and between them and their abiotic environments (Norris, 1986, 1988, 1994a,b; Neupane and Norris, 1992; Norris and Liu, 1992; Markovic et al., 1998).

More about thiol/disulfide redox-dependent transduction of energy as information

The essential qualitative and quantitative conversion of abiotic energy states, as in pheromones and phytochemical messengers, into biotic informational energy states occurs via sulfhydryl/disulfide proteinaceous receptors. These are variously associated with the plasma membrane of living cells (Figure 5.3) or the envelope of viruses (Figure 5.4), or found among the cell contents including the nucleus and DNA (Figure 5.5). This life-rendering conversion occurs when the messenger ion, free radical, or molecule, by directly or indirectly oxidizing sulfhydryls or reducing disulfides in proteins of living cells, allows the folding and unfolding, respectively, of the three-dimensional structure of such macromolecules (Sela et al., 1957; Norris, 1971, 1986). More specifically, through chemical-messenger oxidation of a sulfhydryl ($-$SH, thiol) in each of two molecules of the amino acid cysteine in the receptor protein (an energy transducer), a disulfide ($-$S$-$S$-$) may be formed. Such a disulfide formed in the three-dimensional protein involves a conformational (shape) change: the energy transfer thus allows the protein to become more folded (aggregated) in shape (Figure 5.6). Conversely, through chemical-messenger reduction (breakage) of a disulfide, $-$S$-$S$-$, bridge in the receptor protein, the protein may assume a more open (i.e., unfolded) conformation (Figure 5.6). This messenger-driven, highly reversible thiol/disulfide electrochemical aspect of the folding and unfolding of protein (as in muscles during their contractions and relaxations, or in nerve membranes during impulse generation and decay) uniquely brings both qualitative and quantitative macromolecular motions

Figure 5.3 Thiol/disulfide interconversion is fundamental to the functioning of integral membrane proteins, and depends especially on the cell-surface redox potential and the presence of thiol/disulfide isomerases.

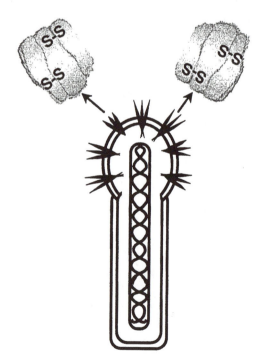

Figure 5.4 Schematic depicting baculovirus envelope of the budded phenotype with protruding gp64 spikes. As shown, gp64 on the envelope exists in the form of two disulfide isomers (Oomens *et al.*, 1995). Possible shuffling between these two gp64 forms with regard to baculovirus infection efficiency is discussed in the text.

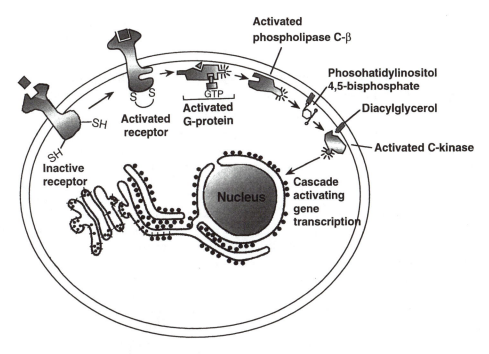

Figure 5.5 An illustration of the signal pathway from the cell surface to the nucleus via G protein-coupled receptors. The finding that *N*-ethylmaleimide (NEM), a thiol-alkylating agent, lowers the receptor affinity (van der Vliet and Bast, 1992) supports the proposal that thiol groups may be critical for formation of the disulfide-dependent conformationally active state of the receptor. (Partially redrawn from Alberts *et al.*, 1994.)

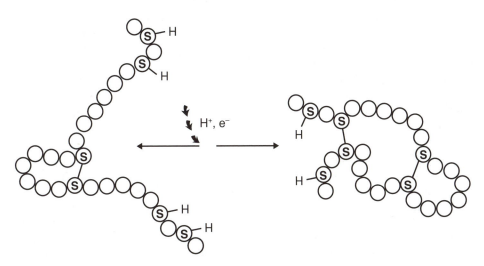

Figure 5.6 Schematic illustrating how the formation of disulfide (—S—S—) bonds in proteins can stabilize the three-dimensional structure of such macromolecules in a more aggregated state (on the right); the reduction of those —S—S— bonds by reducing agents (e.g., some chemical messengers) can yield sulfhydryls (—SH) and impart a more "opened" nonaggregated, three-dimensional form to the protein in the receptor and energy-transducer macromolecule (on the left). In this illustration, each circle in the chains represents an amino acid in the chain of amino acids that make up a protein. Each circle containing a sulfur (S) represents the amino acid cysteine if reduced, and a half-cystine if oxidized.

to such systems. Such traits are responsible for the dynamic forms and functions so critical to life (Harold, 1986).

Transduction of energy in the environment of the plasma membrane of living cells or in the envelope of viruses

Again, the first study of thiol/disulfide redox electrochemistry in the transduction of energy as information in the environment of the cell plasma membrane involved —SH/—S—S— proteinaceous receptors (transducers) in the dendritic membrane of primary insect chemosensitive neurons, and was conducted by Norris and associates as cited previously. The proteinaceous receptors involved are members of what has since become a large family (class) of discovered proteins now known as G protein-coupled receptors. These many thiol/disulfide interchange-dependent receptors are now widely considered as critical to a vast array of signals that are transduced at the cell plasma membrane (van der Vliet and Bast, 1992). In the G protein-coupled receptors, the specific essential coupling is via a thiol group (Figure 5.5).

In addition to the thiol/disulfide-dependent signal transductions in the environment of the cell plasma membrane that are involved in various extracellular–cellular information transfers, another enormously important involvement of thiol–disulfide redox interchange at the cell membrane is in controlling virus penetration of the host cell (Markovic *et al.*, 1998). Such redox-based regulation is exerted at the fusion of the viral envelope with the plasma membrane of the target (host) cell, and in the subsequent delivery of the viral core into the cytoplasm. A well-described specific example is baculovirus and its insect host cells. The envelope of baculovirus contains, among others, a trimeric glycoprotein (gp64), which mediates fusion of the viral envelope with the membrane of the target insect cell and allows entry of the viral capsid into the cell (Blissard and Wenz, 1992). Markovic *et al.* (1998) first identified and characterized a large macromolecular aggregate of gp64 as an essential intermediate in this fusion. The formation of such a macromolecular fusion intermediate is critically dependent upon disulfide bonds connecting monomers into a stable trimer. Reduction of these disulfides inhibits both fusion and formation of the macromolecular gp64 structure (Markovic *et al.*, 1998).

It has been hypothesized that another regulatory role of thiol/disulfide interchange in baculovirus is to drive interconversion of the two disulfide isomers of the gp64 trimer. It is not known whether these two trimeric gp64 forms differ in functional properties; however, they apparently differ in their pattern of disulfide bonding (Oomens *et al.*, 1995). Markovic *et al.* (1998), however, did find a strong correlation between DTT-dependent destabilization of one trimer (trimer 1) and inhibition of fusion. If trimer 1 is more involved in baculovirus fusion than trimer 2, then the latter may serve as a reservoir of the fusion-competent trimer. The interconversion between the two isomers may be mediated by thiol/disulfide isomerases on the surface of the insect cell. A protein disulfide isomerase (PDI) has been implicated in the thiol/disulfide isomerization of another enveloped virus, the human immuno-deficiency virus (HIV) (Ryser *et al.*, 1994). With HIV, cell-surface PDI catalyzes thiol/disulfide interchange, yielding reduction of critical disulfides of the fusion protein involved, gp120/gp41. The interchange and reduction jointly confer on the protein conformational activation required for HIV penetration and cell infection. This capability of PDI as a cell-surface redox-regulating mechanism has only recently been recognized in animal systems by other investigators (Mandel *et al.*, 1993; Jiang *et al.*, 1999). This newly recognized role for PDI on the cell surface is particularly interesting because, although previously detected at the surface of mammalian cells, this isomerase is classically considered a chaperone and catalyst of disulfide-bond formation in nascent polypeptide chains during their folding in the endoplasmic reticulum inside the cell.

By using thiol/disulfide-modifying agents, thiol/disulfide shuffling was shown to be essential for other animal viruses (e.g., sindbis and vaccinia virus) to penetrate and infect the target cell (Abell and Brown, 1993; Locker and Griffiths, 1999). Reduction of critical disulfide bridges during sindbis virus penetration of the target cell disrupts the rigid icosahedral viral lattice, which otherwise facilitates both fusion of the viral envelope with the membrane of the host cell and release of the viral genome into the cell (Abell and Brown, 1993). Such reduction of vaccinia virus core protein similarly disrupts the viral structure necessary for delivery of viral cores into the cytoplasm of the host cell (Locker and Griffiths, 1999).

Entry of numerous bacterial (e.g., diphtheria, cholera, *Pseudomonas* exotoxin A) and plant (e.g., ricin) protein toxins into the target cell also depends upon regulation of thiol/disulfide interconversions. Although these toxins exhibit some variation in their mode of action, they all share a common structural motif (Spangler, 1992; Lord and Roberts, 1998). They generally contain a subunit that is responsible for the binding of the toxin to receptors on the plasma membrane of the susceptible cells, and another subunit that is catalytically active and responsible for the toxicity of the protein. The binding and catalytic subunits are connected covalently by a disulfide linkage. Liberation of the catalytic subunit from the binding subunit, once binding of the latter to the target cell has occurred, is essential for translocation of the former subunit through the plasma membrane of the target cell to intracellular sites of toxic action. The exact mechanism by which this required disulfide reduction occurs has not been elucidated; however, it is reasonable to postulate that thiol/disulfide interchange modulates exchange of protons between a cellular donor (e.g., PDI, glutaredoxin, thioredoxin, glutathione) and an acceptor (i.e., toxin subunit).

Transduction of energy among the cell contents including the nucleus and DNA

The presence and capabilities of thiol/disulfide redox electrochemistry for transferring chemical groups and energy as information are universal in intracellular environments. Considering first the transitional environment from the cell plasma membrane to the cell cytosol, transduction of energy as information via various membrane-channel and gating processes, e.g. Ca^{2+}-driven signaling, is known to depend on the cellular redox potential (Sen, 1998). The naturally occurring intracellular disulfide-reducing agent, reduced glutathione (GSH) (Zabel *et al.*, 1997), can inhibit this. Moving our focus next toward the cell nucleus and the genome in particular, thiol/disulfide interchange is known to regulate gene expression and to control the function of gene-transcriptional factors (Figure 5.5) (e.g., Abate *et al.*, 1990; Storz *et al.*, 1990; Kumar *et al.*, 1992;

Matthews *et al.*, 1993; Reichard, 1993; Paget *et al.*, 1998; Sen, 1998; Zheng *et al.*, 1998). Examples of such altered regulation of gene transcription (Figure 5.5) include mislocalization of periplasmic proteins (Stewart *et al.*, 1998) and changed cellular signaling, cell proliferation, and cell adhesion (Sen, 1998). Sulfhydryl/disulfide exchange further regulates transcription of genes coding for cellular antioxidative (i.e., stress-defensive) chemistry (Stewart *et al.*, 1998). The peptide GSH is among the more important of such reductive, defensive chemicals in plant or animal cells. Zheng *et al.* (1998) found that reversible disulfide bonding of cytoplasmic proteins can be used specifically by the cell as a very sophisticated control of the "on–off" switch that regulates the central signal-transduction pathway in cells (Figure 5.5). It is considered (Zheng *et al.*, 1998) that this major signal-transduction pathway, built centrally around activation–inactivation of the cellular thioredoxin redox system, is critical in the regulation of the expression of genes that confer cell resistance, or susceptibility, to stress factors (e.g., reactive oxidative species or agents). By using the reducing potential of NADPH (the reduced form of nicotinamide–adenine dinucleotide phosphate), thioredoxin reductase may maintain thioredoxin 1, in the cell central redox system, in its reduced state, which keeps cytoplasmic proteins in their reduced state. Keeping this central redox system intact is vital to "healthy" cellular existence, but it may easily be altered (mutated) by electron/proton scavenging (or accepting) entities (i.e., stresses) (Stewart *et al.*, 1998).

From the intracellular perspective, inducible (i.e., alterable) plant defensive chemistry is thus the thiol/disulfide redox-dependent net chemical result of the plant cell's effort to maintain its life-essential central redox system and associated signal-transduction pathway in functional states when challenged by stresses originating from the environment (e.g., insect herbivores).

Conclusions

The existence of plant genomic-dependent alterable extrinsic as well as intrinsic chemical defenses (messages) has been documented. The extrinsic and intrinsic defenses (chemical communications) were shown to be not only compatible but also interdependent. Not only are alterable plant chemical defenses multicomponent but any given component likely serves numerous communicative roles among the diverse array of living forms. This alterable chemistry exists in at least three regimes: first, as volatile repellents; second, as less-volatile feeding deterrents; and third, as inhibitors of herbivore general physiology and development. The example of an almost ideal extreme biofacies (Hamilton, 1967) in the polytrophic *X. ferrugineus* living system clearly demonstrates that elaboration of polytrophism may render diverse plant chemical defenses ultimately quite ineffective. Finally, the findings reported here add compelling support to the interpretation that the transfer of chemical groups and energy as extracellular and intracellular information by most, if not all, living entities is dependent above all upon thiol/disulfide redox chemistry.

Acknowledgments

Research reported here from the first author's laboratories was supported partially by the College of Agricultural and Life Sciences, University of Wisconsin, Madison, WI, USA; in part by numerous research grants from the US National Science Foundation and the US National Institutes of Health; and in part by US Hatch Projects 3040 and 3419, McIntire-Stennis Project 3127, and USDA Competitive Grants 84-CRCR-1-1501 and 88-37153-4043. The second author acknowledges her earlier research support as a Fellow, Laboratory of Cellular and Molecular Biophysics, NICHD, US National Institutes of Health.

References

Abate, C., Patel, L., Rauscher, F.J. III and Curran, T. (1990) Redox regulation of Fos and Jun DNA-binding activity *in vitro. Science*, **249**, 1157–1161.

Abell, B.A. and Brown, D.T. (1993) Sindbis virus membrane fusion is mediated by reduction of glycoprotein disulfide bridges at the cell surface. *J. Virol.*, **67**, 5496–5501.

Adkisson, P.L. and Dyck, V.A. (1980) Resistant varieties in pest management systems. In F.G. Maxwell and P.R. Jennings (eds.), *Breeding Plants Resistant to Insects*, Wiley, New York, pp. 233–252.

Akazawa, T., Uritani, I. and Kubota, H. (1960) Isolation of ipomeamarone and two coumarin derivatives from sweet potato roots injured by the weevil, *Cylas formicaris elegantulus. Arch. Biochem. Biophys.*, **88**, 150–156.

Alberts, B., Bray, D., Lewis, J., Raff, M., Roberts, K. and Watson, J.W. (1994) *Molecular Biology of The Cell*, Garland Publishing, pp. 748–749.

Aldrich, J.R., Blum, M.S. and Duffey, S.S. (1976) Male specific natural products in the bug, *Leptoglossus phyllopus*: chemistry and possible function. *J. Insect Physiol.*, **22**, 1201–1206.

Axelrod, J. and Tomchick, R. (1958) Enzymatic *O*-methylation of epinephrine and other catechols. *J. Biol. Chem.*, **233**, 702.

Baker, J.M. and Norris, D.M. (1968) A complex of fungi mutualistically involved in the nutrition of the ambrosia beetle *Xyleborus ferrugineus. J. Invertebr. Pathol.*, **11**, 246–250.

Barbosa, P. (1998) *Conservation Biological Control*, Academic Press, New York.

Blissard, G.W. and Wenz, J.R. (1992) Baculovirus GP64 envelope glycoprotein is sufficient to mediate pH dependent membrane fusion. *J. Virol.*, **66**, 6829–6835.

Burden, B.J. and Norris, D.M. (1994) Ovarian failure induced in *Epilachna varivestis* by a "death-trap" "Davis" variety of *Glycine max. Entomol. Exp. Appl.*, **73**, 183–186.

Chiang, H.S., Norris, D.M., Ciepiela, A., Shapiro, P. and Oosterwyk, A. (1987) Inducible versus constitutive PI 227687 soybean resistance to Mexican bean beetle, *Epilachna varivestis. J. Chem. Ecol.*, **13**, 741–749.

Crewe, R.M. and Blum, M.S. (1972) Alarm pheromones of the Attini: their phylogenetic significance. *J. Insect Physiol.*, **18**, 31–42.

Crewe, R.M., Blum, M.S. and Collingwood, C.A. (1972) Comparative analysis of alarm pheromones in the ant genus *Crematogaster. Comp. Biochem. Physiol.*, **43B**, 703–716.

Cruickshank, I.A.M. and Perrin, D.R. (1965) Studies on phytoalexins. VIII. The effect of some further factors on the formation, stability and localization of pisatin *in vitro. Aust. J. Biol. Sci.*, **18**, 817–828.

Daroogheh, H. and Olagbemiro, T.O. (1982) Linalool in the cotton stainer *Dysdercus superstitiosus* (F.) (Heteroptera: Pyrrhocoridae). *Experientia*, **38**, 421–423.

Dicke, M., van Beek, T.A., Posthumus, M.A., Ben Dom, van Bokhoven, H. and Groot, Ae. De. (1990) Isolation and identification of volatile kairomone that affects acarine predator–prey interactions. *J. Chem. Ecol.*, **16**, 381–396.

Drost, Y.C., Lewis, W.J. and Tumlinson, H.J. (1988) Beneficial arthropod behavior mediated by airborne semiochemicals. V. Influence of rearing method, host-plant, and adult experience on host-searching behavior of *Microplitis croceipes* (Cresson), a larval parasitoid of *Heliothis. J. Chem. Ecol.*, **14**, 1607–1616.

Duffey, S.S. and Blum, M.S. (1977) Phenol and guaiacol: biosynthesis, detoxification, and function in a polydesmid millipede, *Oxidus gracilis. Insect Biochem.*, **7**, 57–65.

Duffield, R.M. and Blum, M.S. (1975) Identification, role and systematic significance of 3-octanone in the carpenter ant, *Camponotus schaefferi* Whr. *Comp. Biochem. Physiol.*, **51B**, 281–282.

Eller, F.J., Tumlinson, J.H. and Lewis, W.J. (1988) Beneficial arthropod behavior mediated by airborne semiochemicals: source of volatiles mediating the host-location flight behavior of *Microplitis croceipes* (Cresson) (Hymenoptera: Braconidae), a parasitoid of *Heliothis zea* (Boddie) (Lepidoptera: Noctuidae). *Environ. Entomol.*, **17**, 745–753.

Elzen, G.W., Williams, H.J. and Vinson, S.B. (1984) Role of diet in host selection of *Heliothis virescens* by parasitoid *Campoletis sonorensis* (Hymenoptera: Ichneumonidae). *J. Chem. Ecol.*, **10**, 1535–1541.

Elzen, G.W., Williams, H.J., Vinson, S.B. and Powell, J.E. (1987) Comparative flight behavior of parasitoids *Campoletis sonorensis* and *Microplitis croceipes*. *Entomol. Exp. Appl.*, **45**, 175–180.

Ferkovich, S.M. and Norris, D.M. (1972) Antennal proteins involved in the neural mechanism of quinone inhibition of insect feeding. *Experientia*, **28**, 978–979.

Frazier, J.L. and Heitz, J.R. (1975) Inhibition of olfaction in the moth *Heliothis virescens* by the sulfhydryl reagent fluorescein mercuric acetate. *Chem. Senses Flavor*, **1**, 271–281.

Getchell, M.L. and Gesteland, R.C. (1972) The chemistry of olfactory reception: stimulus-specific protection from sulfhydryl reagent inhibition. *Proc. Natl. Acad. Sci. USA*, **69**, 1494–1498.

Gibbs, R.D. (1974) *Chemotaxonomy of Flowering Plants*, vol. 1, McGill-Queen's University Press, Montreal, Canada.

Gilbert, B.L., Baker, J.E. and Norris, D.M. (1967) Juglone (5-hydroxy-1,4-naphthoquinone) from *Carya ovata*, a deterrent to feeding by *Scolytus multistriatus*. *J. Insect Physiol.*, **13**, 1453–1459.

Haanstad, J.O. and Norris, D.M. (1992) Altered elm resistance to smaller European elm bark beetle (Coleoptera: Scolytidae) and forest tent caterpillar (Lepidoptera: Lasiocampidae). *J. Econ. Entomol.*, **85**, 172–181.

Hamilton, W.D. (1967) Extraordinary sex ratios. *Science*, **156**, 477–488.

Harold, F.M. (1986) *The Vital Force: A Study of Bioenergetics*, W.H. Freeman, San Francisco, CA.

Horber, E. (1972) Plant resistance to insects. *Agric. Sci. Rev.*, **10**, 1–18.

Jiang, X.-M., Fitzgerald, M., Grant, C.M. and Hogg, P.J. (1999) Redox control of exofacial protein thiol/disulfides by protein disulfide isomerase. *J. Biol. Chem.*, **274**, 2416–2423.

Kennedy, B. (1984) Effect of multilure and its components on parasites of *Scolytus multistriatus* (Coleoptera: Scolytidae). *J. Chem. Ecol.*, **10**, 373–385.

Kogan, M. and Paxton, J. (1983) Natural inducers of plant resistance to insects. In P.A. Hedin (ed.), *Plant Resistance to Insects*, American Chemical Society, Washington, DC, pp. 153–171.

Kumar, S., Rabson, A.B. and Gelinas, C. (1992) The RxxRxRxxC motif conserved in all Rel/kB proteins is essential for the DNA-binding activity and redox regulation of the v-Rel oncoprotein. *Mol. Cell. Biol.*, **12**, 3094–3106.

Liu, S.H., Norris, D.M. and Marti, E. (1988) Behavioral responses of female adult *Trichoplusia ni* to volatiles from soybeans versus a preferred host, lima bean. *Entomol. Exp. Appl.*, **49**, 99–109.

Liu, S.H., Norris, D.M. and Lyne, P. (1989) Volatiles from the foliage of soybean, *Glycine max*, and lima bean, *Phaseolus lunatus*: their behavioral effects on the insects *Trichoplusia ni* and *Epilachna varivestis*. *J. Agric. Food Chem.*, **37**, 496–501.

Liu, S.H., Norris, D.M., Hartwig, E.E. and Xu, M. (1992) Inducible phytoalexins in juvenile soybean genotypes predict soybean looper resistance in the fully developed plants. *Plant Physiol.*, **100**, 1479–1485.

Liu, S.H., Norris, D.M. and Jianyong, L.P. (1993) Peroxidase activity as correlated with stress-inducible insect resistance in *Glycine max*. *Entomol. (Trends Agric. Sci.)*, **1**, 75–84.

Locker, J.K. and Griffiths, G. (1999) An unconventional role for cytoplasmic disulfide bonds in vaccinia virus proteins. *J. Cell Biol.*, **144**, 267–279.

Lord, M.J. and Roberts, L.M. (1998) Toxin entry: retrograde transport through the secretory pathway. *J. Cell Biol.*, **140**, 733–736.

Ma, W.C. (1977) Alterations of chemoreceptor function in armyworm (*Spodoptera exempta*) by a plant-derived sesquiterpenoid and by sulfhydryl reagents. *Physiol. Entomol.*, **2**, 199–297.

Ma, W.C. (1981) Receptor membrane function in olfaction and gustation: implications from modification by reagents and drugs. In D.M. Norris (ed.), *Perception of Behavioral Chemicals*, Elsevier/North-Holland Biomedical Press, Amsterdam, pp. 267–287.

Mandel, R., Ryser, H.J.-P., Ghani, F., Wu, M. and Peak, D. (1993) Inhibition of a reductive function of the plasma membrane by bacitracin and antibodies against protein disulfide-isomerase. *Proc. Natl. Acad. Sci. USA*, **90**, 4112–4116.

Markovic, I., Haanstad, J.O. and Norris, D.M. (1993) Chemical correlates of alpha-tocopherol (vitamin E) altered *Malacosoma disstria* herbivory in *Fraxinus pennsylvanica* var. *subintegerrinia*, green ash. *J. Chem. Ecol.*, **19**, 1205–1217.

Markovic, I., Norris, D.M. and Cekic, M. (1996a) Some chemical bases for gypsy moth, *Lymantria dispar*, larval rejection of green ash, *Fraxinus pennsylvanica*, foliage as food. *J. Chem. Ecol.*, **22**, 2283–2298.

Markovic, I., Norris, D.M., Phillips, J.K. and Webster, F.X. (1996b) Volatiles involved in the non-host rejection of *Fraxinus pennsylvanica* by *Lymantria dispar* larvae. *J. Agric. Food Chem.*, **44**, 929–935.

Markovic, I., Norris, D.M. and Nordheim, E.V. (1997) Gypsy moth (*Lymantria dispar*) larval development and survival to pupation on diet plus extractables from green ash foliage. *Entomol. Exp. Appl.*, **84**, 247–254.

Markovic, I., Pulyaeva, H., Sokoloff, A. and Chernomordik, L.V. (1998) Membrane fusion mediated by baculovirus gp64 involves assembly of stable gp64 trimers into multiprotein aggregates. *J. Cell Biol.*, **143**, 1155–1166.

Matthews, J.R., Wakasugi, N., Virelizier, J.-L., Yodoi, J. and Hay, R.T. (1993) Thioredoxin regulates the DNA binding activity of NF-kB by reduction of a disulfide bond involving cysteine 62. *Nucleic Acids Res.*, **20**, 3821–3830.

Mooser, G. (1976) *N*-substituted maleimide inactivation of the response to taste-cell stimulation. *J. Neurobiol.*, **7**, 457–468.

Mudd, A., Walter, J.H.H. and Corbet, S.A. (1984) Relative kairomonal activities of 2-acylcyclohexane-1,3-diones in eliciting oviposition behavior from parasite *Nemeritis canescens* (Grav.). *J. Chem. Ecol.*, **10**, 1597–1601.

Neupane, F.P. and Norris, D.M. (1990) Iodoacetic acid alteration of soybean resistance to the cabbage looper (Lepidoptera: Noctuidae). *Environ. Entomol.*, **19**, 215–221.

Neupane, F.P. and Norris, D.M. (1991a) Sulfhydryl-reagent alteration of soybean resistance to the cabbage looper, *Trichoplusia ni. Entomol. Exp. Appl.*, **60**, 239–245.

Neupane, F.P. and Norris, D.M. (1991b) alpha-Tocopherol alteration of soybean anti-herbivory to *Trichoplusia ni* larvae. *J. Chem. Ecol.*, **17**, 1941–1951.

Neupane, F.P. and Norris, D.M. (1992) Antioxidant alteration of *Glycine max* (Fabaceae) defensive chemistry: analogy to herbivory elicitation. *Chemoecology*, **3**, 25–32.

Nordlund, D.A. and Lewis, W.J. (1985) Response of females of the Braconid parasitoid *Microplitis demolitor* to frass of larvae of the noctuids, *Heliothis zea* and *Trichoplusia ni* and to 13-methylhentriacontane. *Entomol. Exp. Appl.*, **38**, 109–112.

Nordlund, D.A. and Sauls, C.E. (1981) Kairomones and their use for management of entomophagous insects. XI. Effect of host plants on kairomonal activity of frass from *Heliothis zea* larvae for the parasitoid, *Microplitis croceipes. J. Chem. Ecol.*, **6**, 1057–1061.

Norris, D.M. (1969) Energy transduction mechanism in olfaction and gustation. *Nature*, **222**, 1263–1264.

Norris, D.M. (1971) A hypothesized unifying mechanism in neural function. *Experientia*, **27**, 531–532.

Norris, D.M. (1977) A molecular and submolecular mechanism of insect perception of certain chemical information in their environment. In V. Labeyrie (ed.), *Comportement des Insectes et Milieu Trophique*, Actes du Colloque International, No. 265, Paris, pp. 81–102.

Norris, D.M. (1979) Chemoreception proteins. In T. Narahashi (ed.), *Neurotoxicology of Insecticides and Pheromones*, Plenum Press, New York, pp. 59–77.

Norris, D.M. (1981) Possible unifying principles in energy transduction in the chemical senses. In D.M. Norris (ed.), *Perception of Behavioral Chemicals*, Elsevier/North-Holland Biomedical Press, Amsterdam, pp. 289–306.

Norris, D.M. (1985) Electrochemical parameters of energy transduction between repellent naphthoquinones and lipoprotein receptors in insect neurons. *Bioelectrochem. Bioenerget.*, **14**, 449–456.

Norris, D.M. (1986) Anti-feeding compounds. In G. Haug and G. Hoffman (eds.), *Chemistry of Plant Protection*, vol. 1, Springer-Verlag, Berlin, pp. 97–146.

Norris, D.M. (1988) *Periplaneta americana* perception of phytochemical naphthoquinones as allelochemicals. *J. Chem. Ecol.*, **14**, 1807–1819.

Norris, D.M. (1992) *Xyleborus ambrosia* beetles: a symbiotic ideal extreme biofacies with evolved polyphagous privileges at monophagous prices. *Symbiosis*, **14**, 229–236.

Norris, D.M. (1994a) A redox-based mechanism by which environmental stresses elicit change in plant-defensive chemistry. In W.J. Mattson, P. Niemela and M. Rousi (eds.), *Dynamics of Forest Herbivory: Quest for Pattern and Principle*, USDA, Forest Service, General Technical Report NC–183, Maui, HI, pp. 46–57.

Norris, D.M. (1994b) Phytochemicals as messengers altering behavior. In T. N. Ananthakrishnan (ed.), *Functional Dynamics of Phytophagous Insects*, Oxford and IBH Publishing, New Delhi, India, pp. 33–54.

Norris, D.M. and Baker, J.M. (1969) Nutrition of *Xyleborus ferrugineus*. I. Ethanol in diets as a tunneling (feeding) stimulant. *Ann. Entomol. Soc. Am.*, **62**, 592–594.

Norris, D.M. and Chu, H.M. (1974) Chemosensory mechanism in *Periplaneta americana*: electroantennogram comparison of certain quinone feeding inhibitors. *J. Insect Physiol.*, **20**, 1687–1696.

Norris, D.M. and Kogan, M. (1980) Biochemical and morphological bases of resistance. In F.G. Maxwell and P.R. Jennings (eds.), *Breeding Plants Resistant to Insects*, Wiley, New York, pp. 23–61.

Norris, D.M. and Liu, S.H. (1992) A common chemical mechanism for insect–plant communication. In S.B.J. Menken, J.H. Visser and P. Harrewijn (eds.), *Proc. 8th Int. Symp. on Insect–Plant Relationships*, Kluwer Academic, Dordrecht, pp. 186–187.

Norris, D.M., Baker, J.E., Borg, T.K., Ferkovich, S.M. and Rozental, J.M. (1970a) An energy-transduction mechanism in chemoreception by the bark beetle, *Scolytus multistriatus*. *Contrib. Boyce Thompson Inst.*, **24**, 263–274.

Norris, D.M., Ferkovich, S.M., Rozental, J.M., Baker, J.E. and Borg, T.K. (1970b) Energy transduction: inhibition of cockroach feeding by naphthoquinones. *Science*, **170**, 754–755.

Norris, D.M., Ferkovich, S.M., Baker, J.E., Rozental, J.M. and Borg, T.K. (1971) Energy transduction in quinone inhibition of insect feeding. *J. Insect Physiol.*, **17**, 85–97.

Oomens, A.G.P., Monsma, S.A. and Blissard, G.W. (1995) The baculovirus GP64 envelope fusion protein: synthesis, oligomerization, and processing. *Virology*, **209**, 592–603.

Paget, M.S.B., Kang, J.-G., Roe, J.-H. and Buttner, M.J. (1998) σ^R, an RNA polymerase sigma factor that modulates expression of the thioredoxin system in response to oxidative stress in *Streptomyces coelicolor* A3(2). *EMBO J.*, **17**, 5776–5782.

Raina, A.K., Kingan, T.G. and Mattoo, A.K. (1992) Chemical signals from host plant and sexual behavior in a moth. *Science*, **255**, 592–594.

Ramachandran, R. and Norris, D.M. (1991) Volatiles mediating plant–herbivore–natural enemy interactions: electroantennogram responses of soybean looper, *Pseudoplusia includens*, and a parasitoid, *Microplitis demolitor*, to green leaf volatiles. *J. Chem. Ecol.*, **17**, 1665–1691.

Ramachandran, R., Norris, D.M., Phillips, J.K. and Phillips, T.W. (1991) Volatiles mediating plant–herbivore–natural enemy interactions: soybean looper frass volatiles, 3-octanone and guaiacol, as kairomones for the parasitoid, *Microplitis demolitor*. *J. Agric. Food Chem.*, **39**, 2310–2317.

Reichard, P. (1993) From RNA to DNA, why so many ribonucleotide reductases? *Science*, **260**, 1773–1777.

Rodriquez, E. and Levin, D.A. (1976) Biochemical parallelisms of repellents and attractants in higher plants and arthropods. In J.W. Wallace and R.L. Mansel (eds.), *Biomedical Interaction Between Plants and Insects*. Series: Recent Advances in Phytochemistry, vol. 10, pp. 214–270.

Rozental, J.M. and Norris, D.M. (1973) Chemosensory mechanism in American cockroach olfaction and gustation. *Nature*, **244**, 370–371.

Rozental, J.M. and Norris, D.M. (1975) Genetically variable olfactory receptor sensitivity in *Periplaneta americana*. *Life Sci.*, **17**, 105–110.

Ryan, C.A. (1983) Insect-induced chemical signals regulating natural plant protection responses. In R.F. Denno and M.S. McClure (eds.), *Variable Plants and Herbivores in Natural and Managed Systems*, Academic Press, New York, pp. 43–60.

Ryser, H.J.-P., Levy, E.M., Mandel, R. and DiSciullo, G.J. (1994) Inhibition of human immunodeficiency virus infection by agents that interfere with thiol-disulfide interchange upon virus-receptor interaction. *Proc. Natl. Acad. Sci. USA*, **91**, 4559–4563.

Samuelson, G.A. (1981) A synopsis of Hawaiian *Xyleborini* (Coleoptera: Scolytidae). *Pacific Insects*, **23**, 50–92.

Sauls, C.E., Nordlund, D.A. and Lewis, W.J. (1979) Kairomones and their use for management of entomophagous insects: VIII. Effect of diet on the kairomonal activity of frass from *Heliothis zea* (Boddie) larvae for *Microplitis croceipes* (Cresson). *J. Chem. Ecol.*, **5**, 363–369.

Sela, M., White, F.H. and Anfinsen, C.B. (1957) Reductive cleavage of disulfide bridges in ribonuclease. *Science*, **125**, 691–692.

Sen, C.K. (1998) Redox signaling and the emerging therapeutic potential of thiol antioxidants. *Biochem. Pharmacol.*, **55**, 1747–1758.

Sequeira, L. (1983) Mechanisms of induced resistance in plants. *Annu. Rev. Microbiol.*, **37**, 51–79.

Shimada, I., Shiraishi, A., Kijima, H. and Morita, H. (1972) Effects of sulfhydryl reagents on the labellar sugar receptor of the fleshfly. *J. Insect Physiol.*, **18**, 1845–1855.

Shimada, I., Shiraishi, A., Kijima, H. and Morita, H. (1974) Separation of two receptor sites in a single labellar sugar receptor of the fleshfly by treatment with *p*-chloromercuribenzoate. *J. Insect Physiol.*, **20**, 605–621.

Singer, G., Rozental, J.M. and Norris, D.M. (1975a) Sulfhydryl groups and the quinone receptor in olfaction and gustation. *Nature*, **256**, 222–223.

Singer, G., Rozental, J.M. and Norris, D.M. (1975b) ^{35}Sulfur incorporation in neural membrane protein involved in *Periplaneta americana* chemoreception. *Experientia*, **30**, 1384–1385.

Spangler, B.D. (1992) Structure and function of cholera toxin and the related *Escherichia coli* heat-labile enterotoxin. *Microbiol. Rev.*, **56**, 622–647.

Stewart, E.J., Aslund, F. and Beckwith, J. (1998) Disulfide bond formation in the *Escherichia coli* cytoplasm: an *in vivo* reversal for the thioredoxins. *EMBO J.*, **17**, 5543–5550.

Storz, G., Tartaglia, L.A. and Ames, B.N. (1990) Transcriptional regulator of oxidative stress-inducible genes: direct activation by oxidation. *Science*, **248**, 189–194.

Stossel, A. (1984) Regulation by sulfhydryl groups of glyceollin accumulation in soybean hypocotyls. *Planta*, **160**, 314–319.

Szent-Gyorgyi, A. (1973) The development of bioenergetics. In J. Avery (ed.), *Membrane Structure and Mechanisms of Biological Energy Transduction*, Plenum Press, New York, pp. 1–4.

Takabayashi, J. and Takahashi, S. (1989) Effects of fecal pellet and synthetic kairomone on host-searching and postoviposition behavior of *Apanteles kariyai*, a parasitoid of *Pseudaletia separata*. *Entomol. Exp. Appl.*, **52**, 221–227.

Turlings, T.C.J., Tumlinson, J.H. and Lewis, W.J. (1990) Exploitation of herbivore-induced plant odors by host-seeking parasitic wasps. *Science*, **250**, 1251–1253.

Turlings, T.C.J., Tumlinson, J.H., Eller, F.J. and Lewis, W.J. (1991) Larval-damaged plants: source of volatile synomones that guide the parasitoid *Cotesia marginiventris* to the micro-habitat of its host. *Entomol. Exp. Appl.*, **58**, 75–82.

van Alphen, J.J.M. and Vet, L.E.M. (1986) An evolutionary approach to host finding and selection. In J.K. Waage and D.J. Greathead (eds.), *Insect Parasitoids*, Academic Press, London, pp. 23–61.

Vande Berg, J.S. (1981) Ultrastructural and cytochemical parameters of chemical perception. In D.M. Norris (ed.), *Perception of Behavioral Chemicals*, Elsevier/North-Holland Biomedical Press, Amsterdam, pp. 103–131.

van der Vliet, A. and Bast, A. (1992) Effect of oxidative stress on receptors and signal transmission. *Chem.-Biol. Interact.*, **85**, 95–116.

Villet, R.H. (1974) Involvement of amino and sulfhydryl groups in olfaction transduction in silk moths. *Nature*, **248**, 707–708.

Vinson, S.B. (1976) Host selection by insect parasitoids. *Annu. Rev. Entomol.*, **21**, 109–133.

Vinson, S.B. (1981) Habitat location. In D.A. Nordlund, R.L. Jones and W.J. Lewis (eds.), *Semiochemicals: Their Role in Pest Control*, Wiley, New York, pp. 51–77.

Wald, G. (1969) Life in the second and third periods; or why phosphorus and sulfur for high-energy bonds?. In H.M. Kalckar (ed.), *Biological Phosphorylations: Development of Concepts*, Prentice-Hall, New York, pp. 156–167.

Weseloh, R.M. (1981) Host location by parasitoids. In D.A. Nordland, R.l. Jones and W.J. Lewis (eds.), *Semiochemicals : Their Role in Pest Control*, Wiley, New York, pp. 79–95.

Wood, S.I. (1982) The bark and ambrosia beetles of North and Central America (Coleoptera: Scolytidae), a taxonomic monograph. *Great Basin Naturalist Memoirs*, **6**, 1–1359.

Zabel, A.C., Favero, T.G. and Abramson, J.J. (1997) Glutathione modulates ryanodine receptor from skeletal muscle sarcoplasmic reticulum. *J. Biol. Chem.*, **272**, 7069–7077.

Zheng, M., Aslund, F. and Storz, G. (1998) Activation of the OxyR transcription factor by reversible disulfide bond formation. *Science*, **279**, 1718–1721.

Influence of Plant Diversity on Herbivores and Natural Enemies

6

John E. Banks
Environmental Science, Interdisciplinary Arts & Sciences, University of Washington, Tacoma, USA

Introduction

Vegetation diversity has long been recognized as a central player in the structure and regulation of animal populations (Elton, 1958). From time-honored agricultural traditions such as intercropping, trap cropping, and strip cropping, farmers and academics alike have learned that manipulating plant diversity within agroecosystems can be an effective component of insect pest control strategies (Risch *et al.*, 1983; Andow, 1991). In recent decades, agroecologists, conservationists, and restoration ecologists have become increasingly aware of the complex population dynamics that may arise from the interplay of spatial dispersion and vegetation diversity of habitat patches (Bach, 1980a; Letourneau, 1987; Lawrence and Bach, 1989). As metapopulation theory and patch dynamics are applied to a growing number of scenarios, resource managers and researchers are moving beyond simply exploring the patterns of insect distributions and are emphasizing and elucidating underlying mechanisms governing those distributions. Leading the way in these explorations are field experiments and mathematical models that attempt to integrate insect behavior across insect/plant taxa and scales (Harrison, 1989; Schultz, 1998; Schultz and Crone, 2001).

Single-species versus multispecies: trade-offs in tractability

The effects of vegetation diversity on insect populations of herbivores and natural enemies are seemingly as diverse as the respective orders of plant and animal life involved in their interactions. One approach to exploring herbivore–natural enemy interactions is to manipulate vegetation diversity and record the effects on herbivores and natural enemy populations separately. Indeed, much of the pioneering work in invertebrate population dynamics highlights the differential response of predators and prey to the texture of the environment (aquatic or terrestrial) in which they live (Gause, 1934; Huffaker, 1958; Luckinbill, 1975; Kareiva, 1987). More recently, researchers are moving away from single-species analyses and focusing on important and often subtle properties of multi-species interactions (Fagan *et al.*, 1999; Hamback *et al.*, 2000). This is in response to a growing awareness of the importance of the complex interactions that often arise from plant–herbivore–natural enemy systems. Secondary plant chemical compounds often act both as repellent and phagostimulant for herbivores as well as a signal to predators and parasitoids (Bernays and Chapman, 1994), creating a more complex web of interaction (Figure 6.1). Unfortunately, analyses of anything but the simplest multispecies inter-actions often prove to be intractable, a fact of life that is reflected by an overwhelming bias in the literature toward single-species studies manipulations. Despite the inherent simplifications in single-species approaches, much of what we understand about multi-species interactions and community ecology with respect to plant diversity has been gleaned from such studies. Here I discuss herbivores and their natural enemies in turn, focusing on interactions between them, set against a backdrop of vegetation diversity.

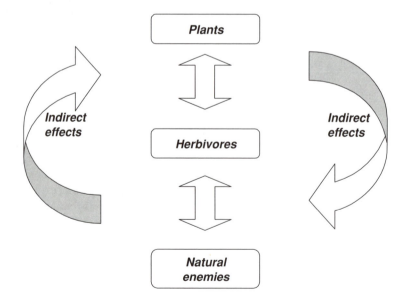

Figure 6.1 Schematic illustrating the relationship between plants, herbivores, and their natural enemies.

John E. Banks

Herbivore distribution and abundance: if you build it, they will come

From the early days of experimental ecology, plant diversity has long been associated with insect population dynamics (Elton, 1958) through debates on the relationship between diversity and stability in ecosystems (May, 1973). Much of the evidence garnered by ecologists through the decades regarding diversity and population dynamics has been in plant-insect systems. Root (1973) and later Vandermeer (1989) proposed hypotheses regarding the reduced ability of herbivores to colonize and remain in diversified plantings due largely to confusing chemical, visual, and olfactory cues. These hypotheses have been well tested during the last two decades, with much evidence suggesting that diversifying vegetation can lead to reductions in herbivore densities (Cromartie, 1975; Bach, 1980b; Banks, 2000) (see Figure 6.2). However, the magnitude and direction of the effects of vegetation diversification on herbivores may be a function of whether the insects are specialists or generalists (Andow, 1983; Sheehan, 1986). Indeed, literature reviews (Risch *et al.*, 1983; Andow, 1991) and more recently meta-analyses (Tonhasca and Byrne, 1994) of diversification studies indicate that, overall, the results of diversification experiments are fairly equivocal, with only a little more than 60 percent of experimental trials resulting in decline of herbivores in one survey (Table 6.1). This is not surprising, given the breadth of insects, plant families, locations, and temporal patterns associated with such studies. Furthermore, interactions between insect dispersal ability and planting spatial scale may be responsible for variable responses to vegetation patterning (Grez and Gonzalez, 1995). In

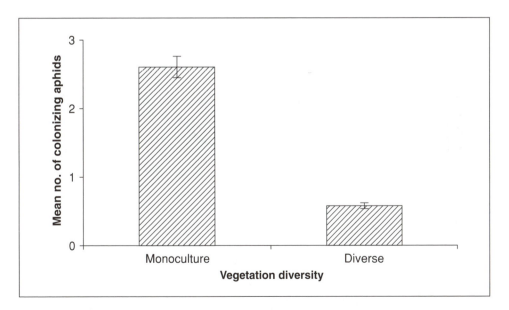

Figure 6.2 Alate aphids colonizing single-species stands of broccoli (monoculture) and mixed plots of broccoli and weedy vegetation (diverse) (After Banks, 2000).

Table 6.1 Responses of herbivores to increases in vegetation diversity relative to control in experiments published in 150 studies

Less abundant	No difference	More abundant
496 cases (62.2%)	207 cases (26.1%)	89 cases (11.2%)

Source: Risch *et al.* (1983)

fact, a recent survey of 17 years of research on how vegetation diversity affects herbivores revealed a marked decrease in the effects of intercropping at large spatial scales (Bommarco and Banks, in review). In common garden experiments, I have found that cabbage aphids, *Brevicoryne brassicae* (Linnaeus), respond to the percentage of habitat comprised by their host plants in the field, but not to the spatial scale at which host-plant patches are separated (Banks, 1998). In contrast, fleabeetles, *Phyllotreta cruciferae* (Goeze), do not respond to vegetation composition per se, but are strongly influenced by the scale at which host-plant patches are deployed in the field, as well as by interaction between vegetation composition and spatial scale (Banks, 1998).

Although ecologists are increasingly interested in how spatial scale affects ecological phenomena (Wiens, 1989; Doak *et al.*, 1992; Underwood and Chapman, 1996; Pascual *et al.*, 2001), scale often receives scant attention in ecological experimental design. As a simple illustration, a survey of herbivore studies referenced in the popular life sciences literature database BIOSIS (Biological Abstracts Inc., Philadelphia, PA, USA) for the past decade underscores the slow pace at which the consideration of spatial scale in published research is progressing (Figure 6.3). Accordingly, an under-

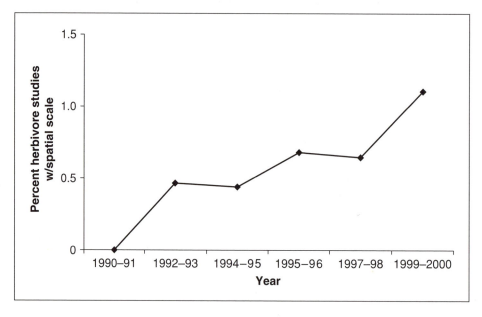

Figure 6.3 Results of a survey of scientific literature in the BIOSIS database from 1990 to 2000. *Y*-axis values represent the percentage of studies referenced by the word "herbivore" that are also referenced by "spatial scale."

John E. Banks

standing of the most influential factors in determining insect responses to diversity remains an obstacle to developing a general protocol in applied endeavors such as intercropping and trap cropping. Progress toward this end will likely increase as more researchers focus on combining literature-based or field-derived parameter values and simple models (Kareiva, 1982, 1985; Banks and Ekbom, 1999).

Diversity and natural enemies: pattern and scale

Apart from interfering with herbivore colonization, more diverse habitats often support more diverse and abundant natural enemies. Field studies have borne this out, which is often due to the fact that more diverse vegetation offers a wider range of resources (e.g., prey items, nectar, pollen, shelter) for predators and parasitoids than do simpler habitats (Harmon *et al.*, 2000). As with herbivores, the effects of diversified vegetation on predators are often scale-dependent (Marino and Landis, 1996; Banks, 1999; Doak, 2000). For instance, in a factorial-design field experiment, I found that *Coccinella septempunctata* Linnaeus is sensitive to the spatial scale at which the host plants of its aphid prey are deployed, whereas the carabid beetle, *Pterostichus melanarius* (Illiger), is not (Banks, 1999). Discrepancies of this sort within a predator complex certainly need to be considered when exploring tritrophic relationships between vegetation diversity, herbivores, and their natural enemies.

Phytochemistry may also mitigate natural enemy–herbivore interactions. For instance, predators may evoke evasive behaviors in herbivores that can indirectly reduce herbivore survivorship by causing them to spend large amounts of time hiding on parts of their host plants that provide only marginal nutrition (Stamp and Bowers, 1991). In other cases, plant defensive chemical compounds may directly alter herbivore behavior and indirectly alter higher trophic levels. In crucifers, for instance, glucosinolates that act as a deterrent to some herbivores are used as a cue for natural enemies of cruciferous pests; recent work has documented the effects of this complex interaction not only on the herbivores but also on the fitness of the predators preying on them (Francis *et al.*, 2001). Vegetation diversity adds yet another complication into this mix. In some cases, the impact of natural enemies on herbivore populations may be different in monocultures than in more diverse habitats — whether this is due to differences in multitrophic responses to phytochemistry, herbivore densities, microclimate, or sensory perception in more or less diverse habitats warrants further exploration (Figure 6.4).

In the past few decades, research has become more focused on testing the precise mechanisms underlying these natural enemy responses to habitat diversity and scale (Kareiva, 1982; Garcia and Altieri, 1992). One approach to such tests is to observe insect movement behavior (which is intrinsically intertwined with habitat spatial scale), which can be quantified and compared across different vegetation types (e.g., simple and diverse) using diffusion models (Turchin, 1991, 1998). In broccoli agroecosystems, for instance, ladybird beetle movement and tenure time in experimental plots are

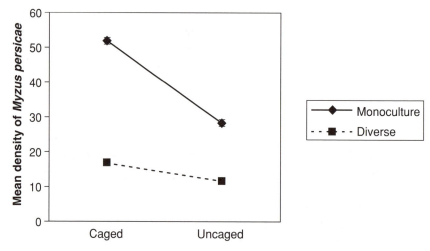

Figure 6.4 Influence of natural enemy exclusion cages on populations of green peach aphids, *Myzus persicae* (Sulzer), in single-species broccoli plots (monoculture) vs. mixed plots (broccoli and weed species) (diverse) (After Banks, 2000). The steeper slope in the monoculture case indicates a stronger effect of predation in less-diverse plots.

sensitive to both plant diversity and the spatial scale at which plants are deployed (Banks, unpublished data). These same methods can be used to predict how natural enemy response might change across a variety of habitats (Firle *et al.*, 1998).

Finally, interactions among predators in a given ecosystem may yield some very interesting dynamics. In some agroecosystems, two different natural enemy species may "cooperate" in suppressing herbivore populations; this type of synergistic interaction has recently been nicely documented in field experiments with aphids, ladybird beetles, and ground beetles (Losey and Denno, 1998). In other cases, natural enemies exert a linearly additive effect on their prey (Chang, 1996), or even interfere with one another to the extent that they have a sub-additive effect on herbivores (Rosenheim *et al.*, 1993). In each of these cases, life history details of both natural enemies and their prey, along with habitat composition and scale, are essential predictors of natural enemy–herbivore interactions. In the past two decades, we have learned much from field experiments aimed at producing mechanistic explanations for herbivore and natural enemy distributions (e.g., Kareiva, 1987). Recently, insights from earlier studies, such as the ability of herbivores to find spatial refuges to escape predation/parasitism, have been brought to bear on better understanding of natural phenomena such as insect outbreaks, with applications in conservation biology and biological control (Kruess and Tscharntke, 1994; Maron *et al.*, 2001).

Genetic engineering, selective pesticides, plant diversity, and insect populations: promise or peril?

As the acreage of crop plants genetically modified to protect crops from insect herbivore attack (*Bt* corn, soybeans, etc.) increases dramatically in the United States, researchers attempt to integrate this new technology with traditional integrated pest management strategies. Furthermore, as the US Environmental Protection Agency continues to implement the Food Quality Protection Act (FQPA) passed by Congress in 1996, agricultural growers and researchers are actively seeking innovative alternatives to traditional broad-spectrum pesticides, many of which have been or are slowly being phased out. Concurrent with the widespread deployment of genetically altered crops in the landscape, then, is an increase in the formulation and deployment of selective, biorational pesticides. These pesticides are designed to target pest insects only, leaving the balance of natural communities intact (but see Stark and Banks, 2001, for some notable exceptions). This increase in both transgenic crops and selective insecticides means renewed hope for combinations of biological control and limited spraying in fields; a stepping-stone that should lead the way for growers to further reduce their reliance on the chemical control of pest insects. Ecological research into the effects of the widespread deployment of genetically modified organisms in the last decade has been concerned primarily with the potential for negative environmental impacts. These researches also include accelerated insect herbivore resistance problems associated with *Bt* crops (Tabashnik, 1994; Peck *et al.*, 1999; Tabashnik *et al.*, 2000) and the escape of transgenes into the environment (Kareiva *et al.*, 1994; Morris *et al.*, 1994). More recently, however, new attention has been focused on the expected impact of integrating transgenes and nontransgenic selective pesticides into diverse crop fields on the management of herbivore populations (Banks and Stark, 1998; Gould, 1998). As we explore these new and complex issues, we hope to gain a better understanding of natural enemies and herbivore responses — an integral part of working to ensure the sustainability and environmental safety of these resources.

Conclusions

As technological advances and environmental crises drive scientific inquiry, research into the interplay of predator–prey relationships with vegetation diversity becomes increasingly important. Just as research into augmentative biocontrol programs in the past stimulated research into multipredator community dynamics, mixtures of selective pesticide use and transgenic plants now inspire further investigation into the effects of multiple disturbances or toxins on both herbivore and natural enemy populations. Finally, as public demand for safer food production drives more growers to implement alternative pest control methods at a variety of scales, ecologists need to focus more on understanding predator and prey behavior at the individual, population, and community levels, explicitly incorporating spatial scale into experimental designs.

Acknowledgment

This is contribution 1004 for the Center for the Study of Invertebrates and Society.

References

Andow, D. (1983) The extent of monoculture and its effects on insect pest populations with particular reference to wheat and cotton. *Agric. Ecosyst. Environ.*, **9**, 25–35.

Andow, D. (1991) Vegetational diversity and arthropod population response. *Annu. Rev. Entomol.*, **36**, 561–586.

Bach, C.E. (1980a) Effects of plant density and diversity on the population dynamics of a specialist herbivore, the striped cucumber beetle *Acalymma vittata* (Fab.). *Ecology*, **61**, 1515–1530.

Bach, C.E. (1980b) Effects of plant diversity and time of colonization on an herbivore–plant interaction. *Oecologia*, **44**, 319–326.

Banks, J.E. (1998) The scale of landscape fragmentation affects herbivore response to vegetation heterogeneity. *Oecologia*, **117**, 239–246.

Banks, J.E. (1999) Differential response of two agroecosystem predators, *Pterostichus melanarius* (Coleoptera: Carabidae) and *Coccinella septempunctata* (Coleoptera: Coccinellidae), to habitat composition and fragmentation scale manipulations. *Can. Entomol.*, **131**, 645–658.

Banks, J.E. (2000) Effects of weedy field margins on *Myzus persicae* (Hemiptera: Aphididae) in a broccoli agroecosystem. *Pan-Pac. Entomol.*, **76**, 95–101.

Banks, J.E. and Ekbom, B. (1999) Modeling herbivore movement and colonization: pest management potential of cropping and intercropping. *Agric. Forest Entomol.*, **1**, 135–140.

Banks, J.E. and Stark, J.D. (1998) What is ecotoxicology? An ad-hoc grab bag or an interdisciplinary science? *Integr. Biol.*, **1**, 195–204.

Bernays, E.A. and Chapman, R.F. (1994) *Host-plant Selection by Phytophagous Insects.* Chapman and Hall, New York.

Bommarco, R. and Banks, J.E. (in review) Scale as spoiler in vegetation diversity experiments: a meta-analysis approach. *Submitted to Oikos.*

Chang, G. (1996) Comparison of single versus multiple species of generalist predators for biological control. *Environ. Entomol.*, **25**, 207–212.

Cromartie, W.J. (1975) The effect of stand size and vegetational background on the colonization of cruciferous plants by herbivorous insects. *J. Appl. Ecol.*, **12**, 517–533.

Doak, P. (2000) The effects of plant dispersion and prey density on parasitism rates in a naturally patchy habitat. *Oecologia*, **122**, 556–567.

Doak, D.F., Marino, P.C. and Kareiva, P.M. (1992) Spatial scale mediates the influence of habitat fragmentation on dispersal success: implications for conservation. *Theor. Population Biol.*, **3**, 315–336.

Elton, C.S. (1958) *The Ecology of Invasions by Animals and Plants.* Methuen, London.

Fagan, W.F., Cantrell, R.S. and Cosner, C. (1999) How habitat edges change species interactions. *Am. Nat.*, **153**, 165–182.

Firle, S., Bommarco, R., Ekbom, B. and Natiello, M. (1998) The influence of movement and resting behavior on the range of three carabid beetles. *Ecology*, **79**, 2113–2122.

Francis, F., Haubruge, E., Hastir, P. and Gaspar, C. (2001) Effect of aphid host plant on development and reproduction of the third trophic level, the predator *Adalia bipunctata* (Coleoptera: Coccinellidae). *Environ. Entomol.*, **30**, 947–952.

Garcia, M.A. and Altieri, M.A. (1992) Explaining differences in flea beetle *Phyllotreta cruciferae* Goeze densities in simple and mixed broccoli cropping systems as a function of individual behavior. *Entomol. Exp. Appl.*, **62**, 201–209.

Gause, G.F. (1934) *The Struggle for Existence*. Macmillan (Hafner Press), New York.

Gould, F. (1998) Sustainability of transgenic insecticidal cultivars: integrating pest genetics and ecology. *Annu. Rev. Entomol.*, **43**, 701–726.

Grez, A.A. and Gonzalez, R.H. (1995) Resource concentration hypothesis: effect of host plant patch size on density of herbivorous insects. *Oecologia*, **103**, 471–474.

Hamback, P.A., Agren, J. and Ericson, L. (2000) Associational resistance: insect damage to purple loosestrife reduced in thickets of sweet gale. *Ecology*, **81**, 1784–1794.

Harrison, S. (1989) Long distance dispersal and colonization in the Bay Checkerspot butterfly *Euphydryas editha bayensis*. *Ecology*, **70**, 1236–1243.

Harmon, J.P., Ives, A.R., Losey, J.E., Olson, A.C. and Rauwald, K.S. (2000) *Coleomegilla maculata* (Coleoptera: Coccinellidae) predation on pea aphids promoted by proximity to dandelions. *Oecologia*, **125**, 543–548.

Huffaker, C.B. (1958) Experimental studies on predation: dispersion factors and predator–prey oscillations. *Hilgardia*, **27**, 343–383.

Kareiva, P. (1982) Experimental and mathematical analyses of herbivore movement: quantifying the influence of plant spacing and quality on forage discrimination. *Ecol. Monogr.*, **52**, 261–282.

Kareiva, P. (1985) Finding and losing host plants by *Phyllotreta*: patch size and surrounding habitat. *Ecology*, **66**, 1809–1816.

Kareiva, P. (1987) Habitat fragmentation and the stability of predator–prey interactions. *Nature*, **326**, 388–390.

Kareiva, P., Morris, W. and Jacobi, C.M. (1994) Studying and managing the risk of cross-fertilization between transgenic crops and wild relatives. *Mol. Ecol.*, **3**, 15–21.

Kruess, A. and Tscharntke, T. (1994) Habitat fragmentation, species loss and biological control. *Science*, **264**, 1581–1584.

Lawrence, W.S. and Bach, C.E. (1989) Chrysomelid beetle movements in relation to host-plant size and surrounding non-host vegetation. *Ecology*, **70**, 1679–1690.

Letourneau, D. (1987) The enemies hypothesis: tritrophic interactions and vegetation diversity in tropical agroecosystems. *Ecology*, **68**, 1616–1622.

Levin, S.A. (1992) The problem of pattern and scale in ecology. *Ecology*, **73**, 1943–1967.

Losey, J.E. and Denno, R.F. (1998) Postive predator–predator interactions: enhanced predation rates and synergistic suppression of aphid populations. *Ecology*, **79**, 2143–2152.

Luckinbill, L.S. (1975) Coexistence in laboratory populations of *Paramecium aurelia* and its predator *Didinium nasutum*. *Ecology*, **54**, 1320–1327.

Marino, P.C. and Landis, D.A. (1996) Effect of landscape structure on parasitoid diversity and parasitism in agroecosystems. *Ecol. Appl.*, **6**, 276–284.

Maron, J.L., Harrison, S. and Greaves, M. (2001) Origin of an insect outbreak: escape in space or time from natural enemies? *Oecologia*, **126**, 595–602.

May, R.M. (1973) *Stability and Complexity in Model Ecosystems*. Princeton University Press, Princeton, NJ.

Morris, W., Kareiva, P. and Raymer, P. (1994) Do barren zones and pollen traps reduce gene escape from transgenic crops? *Ecol. Appl.*, **4**, 157–165.

Pascual, M., Mazzega, P. and Levin, S.A. (2001) Oscillatory dynamics and spatial scale: the role of noise and unresolved pattern. *Ecology*, **82**, 2357–2369.

Peck, S.L., Gould, F. and Ellner, S. (1999) Spread of resistance in spatially extended regions of transgenic cotton: implications for management of *Heliothis virescens* (Lepidoptera: Noctuidae). *J. Econ. Entomol.*, **92**, 1–16.

Risch, S., Andow, D. and Altieri, M.A. (1983) Agroecosystem diversity and pest control: data, tentative conclusions, and new research directions. *Environ. Entomol.*, **12**, 625–629.

Root, R.B. (1973) Organization of a plant–arthropod association in simple and diverse habitats: the fauna of collards (*Brassica oleracea*). *Ecol. Mongr.*, **43**, 95–124.

Rosenheim, J.A., Wilhoit, L.R. and Armer, C.A. (1993) Influence of intraguild predation among generalist insect predators on the suppression of an herbivore population. *Oecologia*, **96**, 439–449.

Schultz, C.B. (1998) Dispersal behavior and its implications for reserve design in a rare Oregon butterfly. *Conserv. Biol.*, **12**, 284–292.

Schultz, C.B. and E.E. Crone (2001) Edge-mediated dispersal behavior in a prairie butterfly. *Ecology*, **82**, 1879–1892.

Sheehan, W. (1986) Response by specialist and generalist natural enemies to agroecosystem diversification: a selective review. *Environ. Entomol.*, **15**, 456–461.

Stamp, N.E. and Bowers, M.D. (1991) Indirect effect on survivorship of caterpillars due to presence of invertebrate predators. *Oecologia*, **88**, 325–330.

Stark, J.D. and Banks, J.E. (2001) "Selective pesticides": are they less hazardous to the environment? *BioScience*, **51**, 980–982.

Tabashnik, B.E. (1994) Evolution of resistance to *Bacillus thuringiensis*. *Annu. Rev. Entomol.*, **39**, 47–79.

Tabashnik, B.E., Patin, A.L., Dennehy, T.L., Liu, Y-B., Carriere, Y., Sims, M.A. and Antilla, L. (2000) Frequency of resistance to *Bacillus thuringiensis* in field populations of pink bollworm. *Proc. Natl. Acad. Sci. USA*, **24**, 12980–12984.

Tonhasca, A. Jr. and Byrne, D.A. (1994) The effects of crop diversification on herbivorous insects: a meta-analysis approach. *Ecol. Entomol.*, **19**, 239–244.

Turchin, P. (1991) Translating foraging movements in heterogeneous environments into the spatial distribution of foragers. *Ecology*, **72**, 1253–1266.

Turchin, P. (1998) *Quantitative Analysis of Movement: Measuring and Modeling Population Redistribution in Animals and Plants*. Sinauer Associates, Sunderland, MA.

Underwood, A.J. and Chapman, M.G. (1996) Scales of spatial patterns of distributions of intertidal invertebrates. *Oecologia*, **107**, 212–224.

Vandermeer, J. (1989) *The Ecology of Intercropping*. Cambridge University Press, Cambridge, UK.

Wiens, J.A. (1989) Spatial scaling in ecology. *Funct. Ecol.*, **3**, 385–397.

Effect of Multiparasitism on the Parasitization Behavior of Insect Parasitoids

7

Ernest J. Harris and Renato C. Bautista
United States Pacific Basin Agricultural Research Center, US Department of Agriculture, Honolulu, USA

Introduction

The risks and detrimental consequences associated with the use of chemical pesticides have been a recurring theme in pest control over the last decade. As we persistently protect our food supply from perennial ravages by obnoxious pests, a compelling desire to address this issue will continue to preoccupy our quest for ecologically nondisruptive control strategies into the next millennium. Thus, we are tasked with a mandate to minimize drastically the use of pesticides and remove complete reliance on pesticides for control of pests as we continue to search for less harmful and environmentally sound tactics.

Biological control is a nonchemical method that utilizes natural enemies (entomophagous insects) for suppression and regulation of insect pest populations. It is a control approach that may be combined with other more compatible procedures in an integrated pest management (IPM) program (Knipling, 1992). As one of several conventional control methods, biological control has always been constantly challenged and put to test. Nevertheless, from the time biological control was recognized and put into practice as early as the fifteenth century (Doutt, 1965), it has made tremendous impact in suppression of a number of pest organisms. Complete or partial successes in the use of natural enemies against insect pests and weeds have been reported and reviewed in detail by Caltagirone and Doutt (1989), Andres and Goeden (1969), DeBach *et al.* (1969), Hussey and Bravenboer (1969), and Embree (1969).

Biological control of insect pests rests on the premise that pest population densities are amenable to control either by introduction and subsequent establishment of imported parasitoids into a new habitat (the classical approach) or by sustained, strategic releases of laboratory-reared parasitoids into the host habitat in order to achieve high parasitoid-to-host ratios (the augmentative or inundative approach) (DeBach, 1965). Another approach worth evaluation is hyperparasitism: secondary parasitoids that develop at the expense of primary parasitoids, thereby representing a highly evolved fourth trophic level (Sullivan and Völkl, 1999). There is substantial evidence available that hyperparasitoid community organization follows the same pattern as primary parasitoid communities. Studies on this topic, though reviewed earlier (Sullivan, 1987), should combine field and laboratory studies and should include cues involved in habitat and host location and take account of the importance of interactions with adult primary parasitoids and with hyperparasitoid predators at the habitat level (Völkl, 1997). Such an analysis will help to better understand individual patterns of resource utilization that are one key factor of hyperparasitoid population dynamics and its role in pest control (Sullivan and Völkl, 1999). Regardless of the approach, however, the question whether to introduce and utilize only one species or multiple species of parasitoids in suppression of target pest insects always comes to mind. Considering the magnitude, rigors, and costs of foreign exploration, this subject has remained a highly contested issue among research biologists and policy makers.

Multiparasitism has been researched extensively over several decades. This paper will highlight the findings and controversies that have contributed to our understanding of this biological phenomenon. In addition, we will examine the implication of multiparasitism within the context of single or multiparasitoid introduction in biological control of insect pests.

Nature of parasitoid competition

An understanding of how different species of parasitoids behave and interact in a complex faunal community is contingent upon the formulation of field strategies for parasitoid release. A common interaction in a trophic pyramid is the competition that occurs between or among different parasitoids that exploit the same host resources (Price, 1984). Adult parasitoids may compete for same food supply in the natural habitat, e.g., plant pollens, nectars, or honeydew secreted by sedentary insects such as coccids. However, the competition known as multiparasitism (also referred to as heterospecific or interspecific parasitism), becomes extremely critical when the immatures of different parasitoid species utilize the same host individual as the only medium for nourishment and development (Price, 1984). Nevertheless, competition for the same host individual is not limited only to different species of parasitoids. Scramble (or free-for-all) type of struggle for survival inside a host likewise ensues between or

Ernest J. Harris and Rento C. Bautista

among developing parasitoid immatures of the same species. In this case, the host insect is said to be superparasitized (DeBach, 1965): a host may be oviposited with more than one egg by the same gravid female (self parasitization) or by different females of same species (conspecific parsitization). In retrospect, if parasitoid attack occurs in already parasitized host, it is said to be superparasitized or multiparasitized depending on whether the egg-laying female is of the same or different species. Either way, the immature(s) of the parasitoid that predominates over less-aggressive competitors will successfully complete development inside the host. Gregarious or polyembryonic species of parasitoids will produce several progeny from a cohort of eggs deposited inside individual host, while only one offspring will successfully develop from a host parasitized by strictly solitary endoparasitoids (DeBach, 1965).

Host response behavior of female parasitoids

Orientation to and host finding of target insect

Upon emergence in the natural habitat, female parasitoids must mate, search, locate, and then breed in a host that is suitable for its perpetuation. Doutt (1964) listed and described the major steps that should result in successful parasitization by female parasitoids. These are (a) host habitat location, (b) host location, (c) host acceptance, (d) host suitability, and (e) host regulation. In each of these steps, gravid females respond to and rely on a number of cues perceived from the natural habitat, e.g., shape, color, texture, scents that may be associated with food plant of the pest insect, as well as short-range stimuli that emanate from the target insect itself (Vinson, 1976; Willis, 1997). These stimuli may be visual, olfactory, gustatory, or chemosensory in nature. A number of parasitoids are general host seekers, while others are specific strategists that locate and home in to hosts through very specific cues (Willis, 1997). In fact, akin to some parasitoids are very unique host–parasitoid relationships. For example, fermentation odors caused by fungal decaying microorganisms in infested fruits trigger an oviposition response from gravid *Diachasmimorpha longicaudata*, a larval parasitoid of tephritid fruit fly (Madden, 1968; Greany *et al.*, 1977). In a related study, oviposition response by *Psyttalia fletcheri*, a parasitoid of the melon fly larva, *Bactrocera cucurbitae* (Coquillett), was elicited by specific vibrations or sounds that accompany the movement and feeding activity of host larvae inside infested fruits (Nishida, 1956).

During the host selection process, a gravid female is saddled with the burden of making the right choice to ensure that her resources and energy for reproduction are judiciously allocated. Thus, the steps involved in the scrutiny and evaluation of a host require greater input of energy from a female parasitoid in order that she may determine the suitability and status of a potential host. Apparently, it is of utmost benefit to a gravid female to be the first and only recipient of a host. Where hosts are scarce and patchy in distribution, the earlier an undesirable host can be identified, the earlier a parasitoid can commence a new search (Willis, 1997).

In light of this rationale, the ability of female parasitoids to discriminate between healthy and previously parasitized hosts with accuracy and speed becomes a valuable asset in a highly competitive natural environment. In the event that a parasitoid fails in its attempt to make proper host identification, and hence reject a previously parasitized host, the time and energy spent in previously searched environments are wasted. This is particularly critical in the case of solitary parasitoids. Because only one progeny will survive from a parasitized host, repetitive attacks of same host are disadvantageous and a waste of a parasitoid's time and resources.

Host recognition through marking cues

Host discrimination is a common behavioral trait in hymenoptera (Vinson, 1976). The ability of parasitoids to distinguish between healthy or previously parasitized hosts has been the subject of much inquiry in the behavioral area. To date, experimental findings have provided overwhelming evidence that parasitoids are, in fact, capable of recognizing and identifying through cues hosts that have been parasitized by conspecifics or by multiple species of parasitoids.

Salt (1937) was the first to show, in *Trichogramma evanescens* Westwood, that a marking pheromone, left in the host by a female parasitoid that first lays eggs, served as an inhibitory factor against further attack by other parasitoids. Lloyd (1938) demonstrated a similar phenomenon in *Ooencyrtus kuvanae* (Howard). He observed that, instead of laying eggs in previously parasitized hosts, female *O. kuvanae* retained her eggs until such a time as healthy hosts were found and became readily available for parasitization. These initial reports intensified further investigations by other workers. Consequently, an array of inhibitory markers was reported at various levels in the parasitoid–host trophic food chain. These were described as spoor odors (Flanders, 1951), trail odors (Price, 1970), search-deterrent substances (Matthews, 1974), deterrent pheromones (Greany and Oatman, 1972), host-marking pheromones (Vinson, 1972), and contact pheromone perceived through gustatory stimuli (Mackhauer *et al.*, 1996).

Insect hosts may be marked in two ways, namely, with external cues that can be perceived by female parasitoids at very close range or with internal factors that can only be detected after female parasitoids have inserted their ovipositors into the hosts (Salt, 1937). External markers are commonly observed in parasitoids (Bakker *et al.*, 1967; Rabb and Bradley, 1970; Greany and Oatman, 1971, 1972). They may consist of scratches on the surface of the host egg (Safavi, 1968) or pheromone secretions associated with a female's accessory glands (Robertson, 1968; Jackson, 1969; Greany and Oatman, 1972) or Dufour's gland, as in some braconids (Vinson and Guillot, 1972) or the ichneumonid *Campoletis sonorensis* (Cameron) (Guillot and Vinson, 1972).

Recognition of previously parasitized hosts by means of internal markers is seldom observed in parasitoids nor as fully understood as that of external markers. Nevertheless, some parasitoids are able to discriminate hosts through these markers to a limited extent (Fisher, 1961a,b; King and Rafai, 1970; Greany and Oatman, 1971, 1972; Griffiths, 1971). Behavioral response of female parasitoids consisted of retracting their ovipositors from the hosts in response to a number of cues, such as fluid secretions or venom injected into the host by a different parasitoid (Jackson, 1969; Wylie, 1971; Guillot and Vinson, 1972; Propp and Morgan, 1983). Sensing of physical or physiological changes that have occurred inside the host body, such as hemolymph that resulted from parasitoid attack (Fisher and Ganesalingam, 1970; King and Rafai, 1970; Townes, 1971), is also an important cue. Apparently, the chemosensory structures in the ovipositor detect the oviposition promoting or deterring substances that emanate from suitable or unsuitable hosts (Greany et al., 1977). Interestingly, for some aquatic parasitoids, internal markers may be the only means by which females discriminate against previously parasitized hosts (Jackson, 1958).

Host discrimination by internal markers may not be as pervasive as by external markers because internal markers can only be perceived by the egg-laying female hours after oviposition by another female (Klomp et al., 1980; Strand et al., 1986). Strand and Vinson (1983) reported that, in response to internal markers, *Telenomus heliothidis* Ashmead rejected hosts one day after they were parasitized by a conspecific. During this period, host necrosis, a latent manifestation of the injected venom, had occurred. Similarly, *T. heliothidis* discriminated against hosts parasitized by a different species of parasitoid, *Trichogramma pretiosum* Riley, in response to a venom-induced necrosis that occurred in the host several hours after initial attack by the latter species. Evidently, although internal markers produced a delayed response from egg-laying parasitoids, the same markers enabled parasitoids to recognize hosts previously attacked either by conspecifics or by different species of parasitoids.

Recognition of parasitized hosts by egg-laying parasitoids through external or internal markers (or both) should impact female's reproductive success vis-à-vis efficiency of host parasitization. Not only would the host discrimination cues ensure dispersal of parasitoids and thus reduce the time parasitoids spent scouting for hosts in previously searched environments (Price, 1972), but it may have greater impact in minimizing repetitive attacks on multiparasitized and superparasitized hosts (Griffiths, 1960; Lewis et al., 1972; Richerson and Borden, 1972).

Inter- and intraspecific discrimination of parasitized hosts

Most behavioral studies showed that female parasitoids could easily recognize hosts that had previously been attacked by conspecifics or by the same females (self parasitized) (van Alphen and Visser, 1990) but less so with multiparasitized hosts (Pijls et al., 1995). In fact, except where recognition of trail and marking pheromones has occurred between unrelated species (Price, 1970) or between cleptoparasitoids and hyperparasitoids that respond to specific odors left by primary parasitoids (Arthur et al., 1964; Spradbery, 1968; Read et al., 1970), avoidance of multiparasitized hosts is very rarely reported (Price, 1981; Strien-van Liempt et al., 1981). van Alphen and Visser (1990) noted that, unlike in hosts parasitized by conspecifics, gravid females do not seem to recognize interspecific markers left behind by initial attackers in multiparasitized hosts. As may be the case in *Aphytis* spp., reciprocal recognition of marking pheromones seemed to occur only between closely related species (Rosen and DeBach, 1979). And while the sibling species of *Asobara tabida* and *A. rufescens*, both recognized each other's marks and avoided oviposition in hosts already parasitized by the other species, both species failed to recognize hosts that were parasitized by either *Asobara spec* or *A. persimillis* (Vet *et al.*, 1984).

This subject may not have been explored exhaustively, but some investigators have offered hypotheses as to why interspecific discrimination is less common than self or conspecific discrimination. Vinson (1976) suggested that discrimination of multi-parasitized hosts will not be advantageous to superior species that always win the competition since they will avoid hosts previously parasitized by inferior or poorly competitive species. When survival chances are higher in hosts parasitized by another species, i.e., when the presence of another species prevents encapsulation of the superior species' egg (Nishida and Haramoto, 1953), its recognition of interspecific marks and oviposition in parasitized hosts become advantageous to the superior parasitoid. If an inferior parasitoid can recognize cues in the hosts parasitized by superior parasitoids, then its to her advantage to search for and oviposit in healthy hosts to avoid interspecific competition (van Alphen and Visser, 1990).

Knipling (1992) surmised that the ability or lack of ability of a parasitoid to discriminate between healthy and parasitized hosts is not a factor of major importance for the reproductive success of a parasitoid if the average rate of parasitism it causes is marginal. Following this line of argument, the same idea could be one of the reasons why very few studies (Bakker *et al.*, 1967, 1972; Griffiths, 1971; Diamond, 1974) were devoted to developing ecological models that take host discrimination into account (Knipling, 1992). However, Knipling cautioned that host discrimination behavior by parasitoids should not be misconstrued as inconsequential, considering that parasitoids live in highly competitive natural environments. It should be emphasized that multiple attacks of same host would result in complete loss of efficiency of host parasitism (Knipling, 1992).

Mechanisms of competition inside a host body

When a host is attacked by different species of parasitoids, interspecific competition occurs between or among parasitoid immatures inside the host's body. One species will always win in the struggle that ensues, take possession of the host, and successfully develop into adults. However, what may actually take place inside the host during the struggle between parasitoids has remained a subject of perennial contention.

Research interest in the subject began to proliferate as early as the 1900s. One school of thought that evolved was that an inherently superior parasitoid could engage directly and physically destroy its competitor by using its mandibles (Pemberton and Willard, 1918; van Steenburgh and Boyce, 1937; Ullyet, 1943; Simmonds, 1953). Other investigators suggested that the superior parasitoid instead suppresses development of the inferior competitors by physiological inhibition (Pemberton and Willard,

Ernest J. Harris and Rento C. Bautista

1918; Willard and Mason, 1937; van den Bosch and Haramoto, 1953). In the event that neither of the competing parasitoids has any intrinsic advantage over the other, the proponents of this view believed that the sequence of parasitoid oviposition (i.e., which of the competing species lays eggs first into a host) becomes a critical factor in the outcome of competition (Fisher, 1961a). In this case, the older larvae could eliminate the younger larvae (that hatched from later oviposition) by starvation through asphyxiation (Fiske and Thompson, 1909; Tothill, 1922) or by assaulting them with sharp, well-developed mandibles (Thompson, 1923; Willard 1927; Bess, 1936; Simmonds, 1953), or by the condition of the host (i.e., hemolymph) being rendered physiologically unsuitable through injection of an inhibitory substance or venom by the initial attacker to suppress development of later competitors (Timberlake, 1912; Spencer, 1926; Thompson and Parker, 1930; van den Bosch and Haramoto, 1953; Johnson, 1959).

Amidst all these speculations, Fisher (1961b) demonstrated, with ingenious procedures, the exact mechanisms that may occur during interspecific competition between parasitoids in a multiparasitized host. In his study of the interaction of two endoparasitoid species (Ichneumonidae: Ophioninae) competing for the larva of the moth *Ephestia sericarium* Scott (Phycitidae), he provided experimental demonstration of competition between *Nemeritis canescens* Grav. and *Horogenes chrysostictos* Gmelin when the immatures of both were inside the larva of *E. sericarium*. The results of his findings are summarized as follows.

Physical struggle

When eggs of *N. canescens* and *H. chrysostictos* hatched at the same time inside multiparasitized larva of *E. sericarium*, the young larvae moved freely in the hemolymph of the host larva until they met and engaged in struggle. Dissection of the host larvae showed that the larva of either parasitoid species caused the demise of the other by biting the body of the victim with mandibles. Consequently, the wounded larva weakened and became susceptible to encapsulation or phagocytosis by the host. The struggle that followed between competing parasitoid larvae lasted from a few minutes to half an hour, after which the aggressor resumed its activity until completion of its development. According to Fisher, the sequence of oviposition between parasitoids and the age of competing parasitoid larvae were the two most critical factors that would ensure which species would have the physical advantage during the ensuing competition inside the host's body.

Subsequent studies by other workers confirmed Fisher's observation of the physical struggle that occurred between parasitoids in multiparasitized hosts. Vinson (1972) demonstrated that newly hatched supernumeraries of the parasitoid *Cardiochiles nigriceps* Viereck inside a superparasitized larva of *Heliothis virescens* (Fabricius) used their mandibles to attack other parasitoid larvae that they came into contact with as they wandered freely within the host's hemocoel. Eventually, the loser died and became encapsulated by the host phagocytes. Likewise, Mellini and Baronio (1971) observed the same events in *Macquartia chalconota*, where first instar planidia armed with buccal hooks injured their competitors during physical struggle.

Physiological inhibition

Fisher demonstrated this mechanism by ligaturing the body of the host larva with a cotton thread in order to prevent direct physical contact between the larvae of two competing parasitoids, ensuring that the ligature was loose enough to facilitate free circulation of the hemolymph throughout the host body. Employing a technique developed by Salt (1955, 1957), parasitoid eggs were injected into opposite ends of the host body. Injection of parasitoid eggs was staggered so that the larva of one species was already in the early stages of development before egg hatch of the other species. Fisher noted that an inhibitory substance that was released into the blood of the host by the older larva arrested the growth and development of the much younger larva of the other species. Eventually, the latter died and was eliminated completely from the competition.

Based on earlier suggestions, namely, toxic factor, death by starvation, or metabolic secretions (by older larvae) of salivary, excretory or respiratory products (Timberlake, 1910; Tothill, 1922; Spencer, 1926; Simmonds, 1943), Fisher's study through series of tests failed to confirm the nature of inhibition in younger larvae. Nevertheless, Fisher pointed out that the low oxygen content of host hemolymph was probably the cause of physiological suppression. He supported this hypothesis with his finding that a gross change in the oxygen content of the host larva produced an effect on the outcome of parasitoid competition between *N. canescens* and *H. chrysostictos*. Moreover, because the competing parasitoid larvae did not come into contact with the host's tracheal system, he raised the possibility that the parasitoid larvae obtained oxygen directly from the host's hemolymph. Consequently, a change in the oxygen or carbon dioxide content of the host blood brought about by increased metabolic requirements of the older larvae would in effect physiologically suppress the growth and survival of the much younger larval competitor. In a subsequent study, Fisher (1963) reported that the capacity for survival under conditions of low oxygen tension was minimal for eggs and newly hatched larvae of *N. canescens* but rapidly increased with age. King *et al.* (1976) reported that the cause of death of some competing larvae of *Lixophaga diatraeae* (Townsend) in the sugarcane borer appeared to be anoxia when an excessive number of larvae were attached to the host trachea.

The nature of respiratory inhibition advanced by Fisher shed light on earlier speculations by several authors. While van den Bosch and Haramoto (1953) attributed the granulated appearance of parasitoid immatures in multiparasitized larvae of tephritid fruit flies to cytolyzing action by specific inhibitor, Fisher pointed out that these symptoms were, in fact, products of cytolysis that occurred in parasitoid immatures that eventually died because of prolonged oxygen deficiency in the host hemolymph. Similarly, metabolic inhibition rather than poisoning by toxic secretion could explain why development of younger larvae of *Glypta fumiferanae* (Viereck) was suppressed more than that of *Apanteles fumiferanae* (Viereck) larvae when the latter were deposited much earlier than the former as eggs in spruce budworm larvae (Lewis, 1960). Apparently, an increase in oxygen requirements of much older *Apanteles* larvae resulted in oxygen deprivation as well as growth inhibition of the younger *Glypta* larvae. Moreover, that dipterous parasitoids ceased to grow until sufficient air was provided

Ernest J. Harris and Rento C. Bautista

further strengthens the mechanism of respiratory inhibition postulated by Fisher in his study.

Available information to date indicates that physical attack is very common in mandibulate hymenoptera. The nature of physiological suppression in later instars of parasitoid larvae such as anoxia, changes in nutrient availability, toxins, or catolytic enzymes may likewise be critical in the inhibition process (Vinson and Iwantsch, 1980). However, findings reported on the subject are so diverse that general agreement on the exact mechanisms that govern interspecific competition in multiparasitized host has not yet been reached.

Costs of multiparasitism on progeny survival of competing parasitoids

Exploitation of the same host by different species results in disproportionately larger numbers of progeny produced by a parasitoid species that has the advantage over its competitor. Interaction between parasitoids in a multiparasitized host may be passive or active depending on the oviposition behavior of gravid females and the specific stage of the host attacked (Vinson, 1976). Thus, interaction is passive when different parasitoids attack different stages of the host, namely, an egg versus a larval or pupal endoparasitoid. On the other hand, active interaction occurs when different parasitoids attack the same stage of the host.

Most of the findings that have been reported on the outcome of multiparasitism, with regard to progeny development of competing parasitoids, were generated largely from controlled laboratory experiments. Seldom documented in the field, successful development of one competing parasitoid species over another has been attributed to several factors that may be intrinsic to the parasitoid per se or caused by events that may have enhanced a parasitoid's competitiveness.

The inherent superiority of some parasitoids over their competitors or the timing (sequence) of host attack by competing parasitoids are critical factors that could favor successful survival and development of parasitoid progeny. This was the case in the multiparasitism of a pentatomid stinkbug, *Euschistus conspersus* Uhler, by three scelionid egg parasitoids (Weber *et al.*, 1996). Their data indicated that regardless of whether *Psix tumetanus* oviposited in the host before or after *Trissolcus basalis* (Wollaston), the progeny of *P. tumetanus* always predominated in multiparasitized hosts. Nevertheless, in a treatment combination involving the two color forms of *T. utahensis* or between *T. utahensis* and *T. basalis*, two-thirds of the emerging offspring were of the first-ovipositing parasitoid. Reitz (1996) reported that although the parasitoids *Eucelatoria bryani* and *E. rubentis* both developed in multiparasitized larvae of *Helicoverpa zea* (Boddie), the progeny of *E. bryani* survived better and predominated over *E. rubentis*, again indicating the inherent superiority of the former species over the latter. Bautista and Harris (1997) reported similar findings in the multiparasitism of Oriental fruit fly,

Bactrocera dorsalis (Hendel), by an egg and two larval endoparasitoids. The progeny of the egg parasitoid *Fopius arisanus* predominated over those of the larval parasitoid *Diachasmimorpha longicaudata*. However, active competition for the same host between the larval parasitoids *D. longicaudata* and *Psyttalia incisi* resulted in disproportionately large numbers of progeny by either species depending on which parasitoid the host larvae were exposed to first. Interspecific competition between *Epidinocarsis lopezi* De Santis and *E. diversicornis*, both parasitoids of the cassava mealybug, *Phenacoccus manihoti* Matile-Ferrero, resulted in significantly higher survival rate of *E. lopezi* progeny compared to that of *E. diversicornis* irrespective of which species the hosts were exposed to first (Pijls *et al.*, 1995). However, these authors likewise noted that when hosts were exposed for longer periods, progeny survival of the latter parasitoid increased 2.5-fold. Thus, prolonged duration of host exposure to parasitoids provided *E. diversicornis* additional benefit.

There are situations in which different species of parasitoids seem to tolerate each other, so that, actual competition between different parasitoids hardly takes place within the host. Weseloh (1983) showed that *Apanteles melanoscelus* (Ratzeburg) and *Compsilura concinnata* (Meigen) coexisted in harmony inside multiparasitized larva of the gypsy moth. He explained that there was no direct competition between these parasitoids probably because the immatures of each species filled different niches inside the host's body. Hence, while *A. melanoscelus* exploited the host hemolymph, *C. concinnata* resided exclusively in the host's alimentary canal. Similar observations have been reported in some braconids (Wallner *et al.*, 1982) or between a tachinid and a hymenopterous parasitoid (Miller, 1982). Still, there are instances where the effectiveness or the ability of parasitoids to compete may be dictated primarily by the conditions in the habitat in which they compete. Thus, between the two sibling species of the parasitoid *Aphytis*, while one species was superior at low temperature, the other species performed better at high temperature (DeBach and Gisojeric, 1960). In this case, depending upon seasonal conditions and geographical locality, no single parasitoid species would consistently account for a dominant proportion of the hosts parasitized (Blanchard and Conger, 1932; van den Bosch, 1950).

Competitive displacement: an issue in single or multiparasitoid introduction

Competitive displacement results when interaction between or among different parasitoids brings about the extinction of another species or prevents another species from invading and successfully colonizing all or a part of its habitat (DeBach, 1966). Understandably, parasitoid displacement brought about by interspecific competition affects the abundance, distribution, and common structure of parasitoid guilds in the natural habitat (Huffaker *et al.*, 1972).

Ernest J. Harris and Kento C. Bautista

Pemberton and Willard (1918) first recognized that interspecific competition could potentially be advantageous to the host pest. Their assumption was based on actual field observations of two braconid parasitoids (*Opius* spp.) that were introduced into Hawaii for control of Mediterranean fruit fly, *Ceratitis capitata* (Wiedemann). They noted that the larvae of *O. tryoni* always eliminated *O. humilis* larvae when they occurred together in the same host. On the basis of this observation, they hypothesized that *O. humilis* had the best potential because it had a comparably higher reproductive capacity. They concluded that only the highly fecund parasitoid should have been released into the medfly habitat because the other species tended to negate the reproductive superiority of the other. Another classical case that demonstrated how interspecific competition could disadvantage other potential parasitoids was that of several ectoparasitoids introduced into California for control of citrus red scale, *Aonidiella aurantii* (Maskell) (DeBach, 1962). When *Aphytis chrysomphali* (Mercet) and *Aphytis lingnanensis* Compere attacked the same host, the latter had only very slight reproductive advantage over *A. chrysomphali*. Consequently, *A. lingnanensis* failed to win over another species, *A. melinus* DeBach in the field because of inherent differences in progeny production.

Exclusion or displacement of an inferior but highly reproductive parasitoid by a generally superior but poorly fecund species raised the possibility of decreased total host parasitization (Miller, 1977). Analyses of some empirical data showed that this assumption is not unfounded. For instance, Force (1974) reported that total parasitization of *Rhopalomyia californica* was highest in samples composed of one parasitoid species only but ca. 20 percent lower in samples that consisted of three or more parasitoids. Analysis of interaction among three parasitoids of *Musca domestica* Linnaeus yielded a similar relationship between total host parasitization and number of attacking parasitoids (Ables and Shepard, 1976).

Contrary to above findings, however, there is overwhelming experimental evidence in support of multispecies parasitoid introduction in classical biological control (DeBach, 1966). The majority of published literature indicated that although the effect of one species of parasitoid on a pest population may be reduced by multiparasitism, the total effect of several competing species may be greater than the action of any species alone (Propp and Morgan, 1983). This was the case in a study of multiparasitism of Mediterranean black scale, *Saissetia oleae* (Olivier). Combined parasitization by *Metaphycus lounsburyi* (Howard) and *Scutellista cyanea* (Motschulsky) resulted in higher total host mortality than when either parasitoid acted alone (Ehler, 1978). Likewise, despite some degree of multiparasitism of house fly pupae by *Muscidifurax* spp. and *Spalangia* spp., the total rate of host parasitization was greater when both parasitoid groups were utilized simultaneously, probably because of differences in their searching behavior (Propp and Morgan, 1983). While *Spalangia* spp. attacked buried house fly puparia, *Muscidifurax* spp. preferred to parasitize hosts located on or near the surface of the habitat. An assessment of competitive interactions among three parasitoids of gall-forming midge, *Rhopalomyia californica*, showed that total parasitization in mixed-species samples was greater than parasitization in single-species samples, and that, although interspecific competition occurred among parasitoid guilds, interaction did not result in reduced levels of total parasitization (Ehler, 1979).

Empirical analysis of field data yielded convincing results in favor of multiparasitism. Interspecific competition between hymenopterous parasitoids *S. cyanea* (Pteromalidae) and *M. lounsburyi* (Encyrtidae), natural enemies of Mediterranean black scale, *S. oleae*, showed that percentage parasitization by *M. lounsburyi* was not significantly reduced when in combination with *S. cyanea* (Ehler, 1978). In fact, the total proportion of hosts attacked was greater when both species were present compared to when either of them occurred alone. Similarly, multiple parasitization of yellow-striped armyworm, *Spodoptera praefica*, in hay alfalfa by at least three species of parasitoids did not correspond with a decrease in total host parasitization (Miller, 1977). Moreover, quantitative evaluation of parasitization of cabbage looper larvae, *Trichoplusia ni*, indicated that total parasitization did not decrease with an increase in the number of parasitoid species (Ehler, 1977).

The success or failure of a biological control program depends on how different parasitoids behave and interact with one another (Zwolfer *et al.*, 1976; Ehler and Hall, 1982). Although no serious adversity has ever been documented from previous biological control programs that involved introductions of multiparasitoid species (van den Bosch, 1971), concern has been raised over the issue of multiparasitoid introduction in classical biological control. This is with regard to whether preintroduction studies are needed to determine which parasitoid is the best for introduction or whether to release a cohort of different parasitoids and hope that one species will emerge to be the best (Ehler, 1979). As anticipated, some have endorsed and favored preintroduction studies (Turnbull and Chant, 1961; Watt, 1965; Pschorn-Walcher, 1977) while others have disagreed and questioned the necessity of such approach (Huffaker *et al.*, 1972).

Ehler (1979) and Ehler and Hall (1982) proposed doing preintroductory studies because of the question of interspecific competition. They pointed out that simultaneous release of different parasitoids would encourage host multiparasitism, so that without a preliminary evaluation, the problem of interspecific competition could not be assessed among parasitoids. This may cause reduced levels of overall host parasitization and pest population regulation. Moreover, such studies should improve an understanding of competitive interactions in parasite guilds, thus allowing for more rational decisions as to which parasitoid would be the best competitor (Ehler and Hall, 1982) as well as colonize and release first in the host's natural habitat (Pschorn-Walcher, 1977).

Keller (1984), who disagreed with the rationale of the proponents of the necessity of preintroduction studies, pointed out that multiparasitoid release programs tend to act like "filters" that screen out which species will be the most promising. He contended further that even if different species of parasitoids were released at the same time, competitive interactions would not necessarily occur. He supported this argument with experimental findings by Gross *et al.* (1975, 1981), who demonstrated that some species of entomophagous insects characteristically displayed a tendency to disperse upon release. Keller concluded that Ehler and Hall's analysis was biased because their assumption for allowing competitive exclusion did not ensure competition. Subsequently, Ehler and Hall (1984) asserted their position and argued, that although they do not oppose multispecies introduction, a less empirical approach to such introduction would be more appropriate.

Conclusions

It is evident from various studies that recognition of parasitized hosts by egg-laying parasitoids through markers should impact females' reproducing success vis-à-vis the efficiency of host parasitization. The host discrimination cues ensure dispersal of parasitoids and may have greater impact in minimizing repetitive attacks on multi-parasitized and superparasitized hosts. However, host discrimination behavior by parasitoids should not be misconstrued as inconsequential; in fact, multiple attacks of the same host would result in complete loss of effciency of host parasitism. The studies are so diverse that general agreement on the exact mechanisms that govern interspecific competition in multiparasitized host remains to be reached. Overall successes or failures of parasitoid-based biological control depend on their interaction behavior, and apparently the main concern is to determine whether a specific parasitoid is the best for introduction or whether a cohort of different parasitoids should be released for classical biological control.

Evidence to date is still inadequate to ascertain without question the benefits or detrimental consequences of interspecific competition vis-à-vis competitive exclusion in biological control. Nevertheless, this issue can be put into its proper perspective by quoting a remark made by Huffaker *et al.* (1972), who stated that "natural enemies are not introduced to achieve diversity alone but is done so in an effort to establish a single good species or to find a combination of species that will prove to be better than what is already present."

References

Ables, J.R. and Shepard, M. (1976) Seasonal abundance and activity of indigenous hymenopterous parasitoids attacking the housefly (Diptera: Muscidae). *Can. Entomol.*, **108**, 841–844.

Andres, L.A. and Goeden, R.D. (1969) The biological control of weeds by introduced natural enemies. In C.B. Huffaker (ed.), *Biological Control*, Plenum Press, New York, pp. 143–164.

Arthur, A.P., Stainer, J.E.R. and Turnbull, A.L. (1964) The interaction between *Orgilus obscurator* (Nees) (Hymenoptera: Ichneumonidae), parasites of the pine shoot moth, *Rhycionia buoliana* (Schiff.) (Lepidoptera: Olethreutidae). *Can. Entomol.*, **96**, 1030–1034.

Bakker, K., Bagchee, S.N., van Swet, W.R. and Meelis, E. (1967) Host discrimination in *Pseudeucoila bochei. Entomol. Exp. Appl.*, **10**, 295–311.

Bakker, K., Eijsackers, H.J.P., van Lenteren, J.C. and Meelis, E. (1972) Some models describing the distribution of eggs of the parasite *Pseudeucoila bochei* over its hosts, larvae of *Drosophila melanogaster. Oecologia*, **10**, 29–58.

Bautista, R.C. and Harris, E.J. (1997) Effects of multiparasitism on the parasitization behavior and progeny development of oriental fruit fly parasitoids (Hymenoptera: Braconidae). *J. Econ. Entomol.*, **90**, 757–764.

Bess, H.A. (1936) Biology of *Leschenaultia exul* Townsend, a tachinid parasite of *Malacosoma americana* Fab. and *M. disstria* Hubner. *Ann. Entomol. Soc. Am.*, **29**, 593–613.

Blanchard, R.A. and Conger, C.B. (1932) Notes on *Prodenia praefica* Grote. *J. Econ. Entomol.*, **25**, 1059–1070.

Caltagirone, L.E. and Doutt, R.L. (1989) The history of the vedalia beetle importation to California and its impact on the development of biological control. *Annu. Rev. Entomol.*, **34**, 1–16.

DeBach, P. (1962) Ecological adaptation of parasites and competition between parasitoid species in relation to establishment and success. *Proc. Int. Congr. Entomol.*, **11**, 686–690.

DeBach, P. (1965) The scope of biological control. In P. DeBach (ed.), *Biological Control of Insect Pests and Weeds*, Reinhold, New York, pp. 3–20.

DeBach, P. (1966) The competitive displacement and coexistence principles. *Annu. Rev. Entomol.*, **11**, 183–212.

DeBach, P. and Gisojeric, P. (1960) Effects of temperature and competition on the distribution and relative abundance of *Aphytis lingnanensis* and *A. chrysomphali. Ecology*, **41**, 153–160.

DeBach, P., Rosen, D. and Kenett, C.E. (1969) Biological control of coccids by introduced natural enemies. In C.B. Huffaker (ed.), *Biological Control*, Plenum Press, New York, pp. 165–194.

Diamond, P. (1974) The area of discovery of an insect parasite. *J. Theor. Biol.*, **45**, 467–471.

Doutt, R.L. (1964) Biological characteristics of entomophagous adults. In P. DeBach (ed.), *Biological Control of Insect Pests and Weeds*, Chapman and Hall, London, pp. 145–167.

Doutt, R.L. (1965) The historical development of biological control. In P. DeBach (ed.), *Biological Control of Insect Pests and Weeds*, Reinhold, New York, pp. 21–42.

Ehler, L.E. (1977) Parasitization of cabbage looper in California cotton. *Environ. Entomol.*, **6**, 783–784.

Ehler, L.E. (1978) Competition between two natural enemies of Mediterranean black scale on olive. *Environ. Entomol.*, **7**, 521–523.

Ehler, L.E. (1979) Assessing competitive interactions in parasite guilds prior to introduction. *Environ. Entomol.*, **8**, 558–560.

Ehler, L.E. and Hall, W.R. (1982) Evidence for competitive exclusion of introduced natural enemies in biological control. *Environ. Entomol.*, **11**, 1–4.

Ehler, L.E. and Hall, R.W. (1984) Evidence for competitive exclusion of introduced natural enemies in biological control: an addendum. *Environ. Entomol.*, **13**, v–vii.

Embree, D.G. (1969) The biological control of the winter moth in eastern Canada by introduced parasites. In C.B. Huffaker (ed.), *Biological Control*, Plenum Press, New York, pp. 217–226.

Fisher, R.C. (1961a) A study in insect multiparasitism. I. Host selection and oviposition. *J. Exp. Biol.*, **38**, 267–275.

Fisher, R.C. (1961b) A study in insect multiparasitism. II. The mechanism and control of competition for possession of the host. *J. Exp. Biol.*, **38**, 605–628.

Fisher, R.C. (1963) Oxygen requirements and the physiological suppression of supernumerary insect parasitoids. *J. Exp. Biol.*, **40**, 531–540.

Fisher, R.C. and Ganesalingam, V.K. (1970) Changes in the composition of host hemolymph after attack by insect parasitoid. *Nature*, **227**, 191–192.

Fiske, W.F. and Thompson, W.R. (1909) Notes on the parasites of the Saturniidae. *J. Econ. Entomol.*, **2**, 450–460.

Flanders, S.E. (1951) Mass culture of California red scale and its golden chalcid parasites. *Hilgardia*, **21**, 1–42.

Force, D.C. (1974) Ecology of insect host–parasitoid communities. *Science*, **184**, 624–632.

Greany, P.D. and Oatman, E.R. (1971) Demonstration of host discrimination in the parasite *Orgilus lepidus. Ann. Entomol. Soc. Am.*, **65**, 375–376.

Greany, P.D. and Oatman, E.R. (1972) Analysis of host discrimination in the parasite *Orgilus lepidus. Ann. Entomol. Soc. Am.*, **65**, 377–383.

Greany, P.D., Tumlinson, J.H., Chambers, D.L. and Boush, G.M. (1977) Chemically mediated host finding by *Biosteres (Opius) longicaudatus*, a parasitoid of tephritid fruit fly larvae. *J. Chem. Ecol.*, **3**, 189–195.

Griffiths, D.C. (1960) The behavior and specificity of *Monoctonus paludum* Marshall, a parasite of *Nosonovia ribisnigri* (Mosley) on lettuce. *Bull. Entomol. Res.*, **51**, 303–319.

Griffiths, K.J. (1971) Discrimination between parasitized and unparasitized hosts by *Pleolophus basizonus. Proc. Entomol. Soc. Ont.*, **102**, 83–91.

Gross, H.R. Jr., Lewis, W.J., Jones, R.L. and Nordlund, D.A. (1975) Kairomones and their use for management of entomophagous insects: III. Stimulation of *Trichogramma archaeae, T. pretiosum* and *Microplitis croceipes* with host-seeking stimuli at time of release to improve their efficiency. *J. Chem. Ecol.*, **1**, 431–438.

Gross, H.R. Jr., Lewis, W.J. and Nordlund, D.A. (1981) *Trichogramma pretiosum*: effect of prerelease parasitization experience on retention in release areas and efficiency. *Environ. Entomol.*, **10**, 554–556.

Guillot, F.S. and Vinson, S.B. (1972) Sources of substances which elicit a behavioral response from the insect parasitoid, *Campoletis perdistinctus. Nature*, **235**, 169–170.

Huffaker, C.B., Messenger, P.S. and DeBach, P. (1972) The natural enemy component in natural control and the theory of biological control. In C.B. Huffaker (ed.), *Biological Control*, Plenum, New York, pp. 16–67.

Hussey, N.W. and Bravenboer, L. (1969) Control of pests in glasshouse culture by the introduction of natural enemies. In C.B. Huffaker (ed.), *Biological Control*, Plenum Press, New York, pp. 195–216.

Jackson, D. (1958) Observations on the biology of *Caraphractus cinctus* Walker (Hymenoptera: Myrmaridae), a parasitoid of the eggs of Dytiscidae. I. Methods of rearing and numbers bred on different hosts' eggs. *Trans. Roy. Entomol. Soc. London*, **110**, 533–554.

Jackson, D.J. (1969) Observations on the female reproductive organs and the poison apparatus of *Caraphractus cinctus* Walker. *Zool. J. Linn. Soc.*, **48**, 59–81.

Johnson, B. (1959) Effect of parasitization by *Aphidius platensis* Brethes on the developmental physiology of its host, *Aphis craccivora* Koch. *Entomol. Exp. Appl.*, **2**, 82–99.

Keller, M.A. (1984) Reassessing evidence for competitive exclusion of introduced natural enemies. *Environ. Entomol.*, **13**, 192–195.

King, P.E. and Rafai, J. (1970) Host discrimination in a gregarious parasitoid, *Nasonia vitripennis* (Walker). *J. Exp. Biol.*, **43**, 245–254.

King, E.G., Miles, L.R. and Martin, D.F. (1976) Some effects of superparasitism by *Lixophaga diatraeae* of sugarcane borer larvae in the laboratory. *Entomol. Exp. Appl.*, **20**, 261–269.

Klomp, H., Teerink, B.J. and Wei Chun Ma (1980) Discrimination between parasitized and unparasitized hosts in the egg parasite *Trichogramma embryophagum* (Hym.: Trichogrammatidae): a matter of learning and forgetting. *Neth. J. Zool.*, **30**, 254–277.

Knipling, E.F. (1992) *Principles of Insect Parasitism Analyzed from New Perspectives*. U.S. Dep. Agric. Handb. No. 693, U.S. Government Printing Office, Washington, DC.

Lewis, F.B. (1960) Factors affecting assessment of parasitization by *Apanteles fumiferanae* (Vier.) and *Glypta fumiferanae* (Vier.) on spruce budworm larvae. *Can. Entomol.*, **92**, 881–981.

Lewis, W.J., Sparks, A.N., Jones, R.L. and Barras, D.J. (1972) Efficiency of *Cardiochiles nigriceps* as a parasite of *Heliothis virescens* on cotton. *Environ. Entomol.*, **1**, 468–471.

Lloyd, D.C. (1938) A study of some factors governing the choice of hosts and distribution of progeny by the chalcid *Ooencyrtus kuvanae* Howard. *Phil. Trans. Roy. Soc. London, Biol. Sci.*, **229**, 275–322.

Mackhauer, M., Michaud, J.P. and Völkl, W. (1996) Host choice by Aphidiid parasitoids (Hymenoptera: Aphidiidae): host recognition, host quality and host value. *Can. Entomol.*, **128**, 959–975.

Madden, J. (1968) Behavioral responses of parasites to symbiotic fungus associated with *Sirex noctilio* F. *Nature*, **218**, 189–190.

Matthews, R.W. (1974) Biology of Braconidae. *Annu. Rev. Entomol.*, **19**, 15–32.

Mellini, E. and Baronio, P. (1971) Superparassitismo sperimentale e competizioni larvali del parrasitoide solitario *Macquartia chalconota* Meig. *Boll. Ist. Entomol.*, Univ. Bologna, **30**, 133–152.

Miller, J.C. (1977) Ecological relationships among parasites of *Spodoptera praefica*. *Environ. Entomol.*, **6**, 581–585.

Miller, J.C. (1982) Life history of insect parasitoids involved in successful multiparasitism. *Oecologia*, **54**, 8–9.

Nishida, T. (1956) An experimental study of the ovipositional behavior of *Opius fletcheri*, a parasite of the melon fly. *Proc. Hawaii. Entomol. Soc.*, **16**, 126–134.

Nishida, T. and Haramoto, F. (1953) Immunity of *Dacus cucurbitae* to attack by certain parasites of *Dacus dorsalis*. *J. Econ. Entomol.*, **46**, 61–64.

Pemberton, C.E. and Willard, H.F. (1918) Interrelations of fruit fly parasites in Hawaii. *J. Agric. Res.*, **12**, 285–296.

Pijls, J.W.A.M., Hofker, D.K., van Staalduinen, M.J. and van Alphen, J.J.M. (1995) Interspecific host discrimination and competition in *Apoanagyrus (Epidinocarsis) lopezi* and *A.(E.) diversicornis*, parasitoids of the cassava mealybug *Phenacoccus manihoti*. *Mededelingen van de Faculteit Landbouwwetenschappen van de Rijksuniversiteit Gent.*, **55**, 405–415.

Price, P.W. (1970) Trail odors: recognition by insects parasitic on cocoons. *Science*, **170**, 546–547.

Price, P.W. (1972) Behavior of the parasitoid *Pleolophus basizonus* in response to changes in host and parasitoid density. *Can. Entomol.*, **104**, 129–140.

Price, P.W. (1981) Semiochemicals in evolutionary time. In D.A. Nordlund, R.L. Jones and W.J. Lewis (eds.), *Semiochemicals, Their Role in Pest Control*, Wiley, New York, pp. 25–79.

Price, P.W. (1984) *Insect Ecology*, Wiley, New York.

Propp, G.D. and Morgan, P.B. (1983) Superparasitism of housefly (*Musca domestica* L.) pupae by *Spalangia endius* Walker (Hymenoptera: Pteromalidae). *Environ. Entomol.*, **12**, 561–566.

Pschorn-Walcher, H. (1977) Biological control of forest insects. *Annu. Rev. Entomol.*, **22**, 1–22.

Rabb, R.L. and Bradley, J.R. (1970) Marking host eggs by *Telenomus sphingis*. *Ann. Entomol. Soc. Am.*, **63**, 1053–1056.

Read, D.P., Feany, P.P. and Root, R.B. (1970) Habitat selection by the aphid parasite *Diaeretiella rapae* (Hymenoptera: Braconidae) and hyperparasite *Charips brassicae* (Hymenoptera: Cynipidae). *Can. Entomol.*, **102**, 1567–1578.

Reitz, S.R. (1996) Interspecific competition between two parasitoids of *Helicoverpa zea*: *Eucelatoria bryani* and *E. rubentis*. *Entomol. Exp. Appl.*, **79**, 227–234.

Richerson, J.V. and Borden, J.H. (1972) Host finding behavior of *Coeloides brunneri*. *Can. Entomol.*, **104**, 1235–1250.

Robertson, P.L. (1968) A morphological and functional study of the venom apparatus in representatives of some major groups of Hymenoptera. *Aust. J. Zool.*, **16**, 133–166.

Rosen, D. and DeBach, P. (1979) *Species of Aphytis of the World (Hymenoptera: Aphilinidae)*, Dr. W. Junk, Den Haag.

Safavi, M. (1968) Etude biologique et écologique des hymenopteres parasites des oeufs des punaises des céréales. *Entomophaga*, **13**, 381–495.

Salt, G. (1937) Experimental studies in insect parasitism V. The sense used by *Trichogramma* to distinguish between parasitized and unparasitized hosts. *Proc. Roy. Soc. London*, **122**, 57–75.

Salt, G. (1955) Experimental studies in insect parasitism. VIII. Host reactions following artificial parasitization. *Proc. Roy. Soc. B*, **144**, 380–398.

Salt, G. (1957) Experimental studies in insect parasitism. X. The reactions of some endopterygote insects to an alien parasite. *Proc. Roy. Soc. B*, **147**, 167–184.

Simmonds, F.J. (1943) Superparasitism in *Nemeritis*. *Rev. Canad. Biol.*, **2**, 15–48.

Simmonds, F.J. (1953) Inter-relationships of frit-fly parasites in Eastern North America. *Bull. Entomol. Res.*, **44**, 387–393.

Spencer, H. (1926) Biology of parasites and hyperparasites of aphids. *Ann. Entomol. Soc. Am.*, **19**, 119–157.

Spradbery, J.P. (1968) The biology of *Pseudorhyssa sternata* Merril (Hymenoptera: Ichneumonidae), a cleptoparasite of Siricid woodwasps. *Bull. Entomol. Res.*, **59**, 291–297.

Strand, M.R. and Vinson, S.B. (1983) Factors affecting host recognition and acceptance in the egg parasitoid *Telenomus heliothidis* (Hymenoptera: Scelionidae). *Environ. Entomol.*, **12**, 1114–1119.

Strand, M.R., Meola, S.M. and Vinson, S.B. (1986) Correlating pathological symptoms in *Heliothis virescens* eggs with development of the parasitoid *Telenomus heliothidis*. *J. Insect Physiol.*, **32**, 389–402.

Strien-van Liempt, W.T.F.H. and van Alphen, J.J.M. (1981) The absence of interspecific host discrimination in *Asobara tabida* Nees and *Leptopilina heterotoma* (Thompson), coexisting larval parasitoids of *Drosophila* species. *Neth. J. Zool.*, **31**, 701–712.

Sullivan, D.J. (1987) Insect hyperparasitism. *Annu. Rev. Entomol.*, **32**, 49–70.

Sullivan, D.J. and Völkl, W. (1999) Hyperparasitism: multitrophic ecology and behavior. *Annu. Rev. Entomol.*, **44**, 291–315.

Thompson, W.R. (1923) Recherches sur la biologie des Dipteres parasites. *Bull. Biol.*, **57**, 174–273.

Thompson, W.N. and Parker, H.L. (1930) The morphology and biology of *Enlimneria crossifernur*, an important parasite of the European corn borer. *J. Agric. Res.*, **40**, 321–345.

Timberlake, P.H. (1910) Observations on the early stages of two aphidiine parasites of aphids. *Psyche*, **17**, 125–130.

Timberlake, P.H. (1912) Experimental parasitism: a study in the biology of *Limnerium validum* Cresson. *Bull. U.S. Bur. Entomol.*, **19**, 71–92.

Tothill, J.D. (1922) The natural control of the fall webworm (*Hyphantria cunea* Drury) in Canada. *Tech. Bull. Dom. Can. Dep. Agric.*, **3**, 107 pp.

Townes, H. (1971) Ichneumonidae as biological control agents. *Proc. Tall Timbers Conference on Ecological Animal Control by Habitat Management*, Tall Timbers Research Station, Tallahassee, FL, **3**, 235–248.

Turnbull, A.L. and Chant, D.A. (1961) The practice and theory of biological control of insects in Canada. *Can. J. Zool.*, **39**, 697–753.

Ullyet, G.C. (1943) Some aspects of parasitism in field populations of *Plutella maculipennis* Curt. *J. Entomol. Soc. S. Afr.*, **6**, 65–80.

van Alphen, J.J.M. and Visser, M.E. (1990) Superparasitism as an adaptive strategy for insect parasitoids. *Annu. Rev. Entomol.*, **35**, 59–79.

van den Bosch, R. (1950) The bionomics of *Prodenia praefica* Crote in California. PhD dissertation, University of California, Berkeley, 103 pp.

van den Bosch, R. (1971) Biological control of insects. *Annu. Rev. Ecol. Syst.*, **2**, 45–46.

van den Bosch, R. and Haramoto, F.H. (1953) Competition among parasites of the oriental fruit fly. *Proc. Hawai. Entomol. Soc.*, **15**, 201–206.

van Steenburgh, W.E. and Boyce, H.R. (1937) The simultaneous propagation of *Macrocentrus ancylivorus* Roh. and *Ascogaster carpocapsae* Vier. on the peach moth (*Laspeyresia molesta* Busck.), a study in multiple parasitism. *Rep. Entomol. Soc. Ont.*, **68**, 24–26.

Vet, L.E.M., Meyer, M., Bakker, K. and van Alphen, J.J. (1984) Intra- and interspecific host discrimination in *Asobara* (Hymenoptera) larval endoparasitoids of Drosophilidae: comparison between closely related and less closely related species. *Anim. Behav.*, **32**, 871–874.

Vinson, S.B. (1972) Competition and host discrimination between 2 species of tobacco budworm parasitoids. *Ann. Entomol. Soc. Am.*, **65**, 229–236.

Vinson, S.B. (1976) Host selection by insect parasitoids. *Annu. Rev. Entomol.*, **21**, 109–133.

Vinson, S.B. and Guillot, F.S. (1972) Host-marking: source of a substance that results in host discrimination in insect parasitoids. *Entomophaga*, **17**, 241–245.

Vinson, S.B. and Iwantsch, G.F. (1980) Host suitability for insect parasitoids. *Annu. Rev. Entomol.*, **25**, 397–419.

Völkl, W. (1997) Interactions between ants and aphid parasitoids: patterns and consequences for resource utilization. *Ecol. Stud.*, **130**, 225–240.

Watt, K.F.E. (1965) Community stability and the strategy of biological control. *Can. Entomol.*, **97**, 887–895.

Wallner, W., Weseloh, R.M. and Grinberg, P.S. (1982) Intrinsic competition between *Apanteles melanoscelus* (Hym.: Braconidae) and *Rogas lymantriae* (Hym.: Braconidae) reared on *Lymantria dispar* (Lep.: Lymantriidae). *Entomophaga*, **27**, 99–103.

Weber, C.A., Smilanick, J.M., Ehler, L.E. and Zalom, F.G. (1996) Ovipositional behavior and host discrimination in three scelionid egg parasitoids of stink bugs. *Biol. Cont.*, **6**, 245–252.

Weseloh, R.M. (1983) Effects of multiple parasitism on the gypsy moth parasites *Apanteles melanoscelus* (Hymenoptera: Braconidae) and *Compsilura concinnata* (Diptera: Tachinidae). *Environ. Entomol.*, **12**, 599–602.

Willard, H.F. (1927) Parasites of the pink bollworm in Hawaii. *Tech. Bull. U.S. Dep. Agric.*, No. 19.

Willard, H.F. and Mason, A.C. (1937) Parasitization of the Mediterranean fruit fly in Hawaii, 1914–33. *Circ. U.S. Dep. Agric., No. 109*.

Willis, J.K. (1997) Strategies for successful parasitism. *Insect Behavior Review Articles*, 5 pp.

Wylie, H.G. (1971) Oviposition restraint of *Muscidifurax zaraptor* on parasitized housefly pupae. *Can. Entomol.*, **103**, 1537–1544.

Zwolfer, H., Ghani, M.A. and Rao, V.P. (1976) Foreign exploration and importation of natural enemies. In C.B. Huffaker and P.J. Messenger (eds.), *Theory and Practice of Biological Control*, Academic Press, New York, pp. 189–207.

Synergism Between Insect Pathogens and Entomophagous Insects, and Its Potential to Enhance Biological Control Efficacy

8

S.P. Wraight
USDA-ARS, US Plant, Soil, and Nutrition Laboratory, Ithaca, NY, USA

Introduction

Insect pathogens and entomophagous insects (parasites and predators) possess markedly different modes of insecticidal action and have enormous capacity to interact in ways potentially both detrimental and beneficial in terms of pest control. Interactions between pathogens and entomophagous insects have been a major concern to biological control researchers for many years and have been the focus of numerous reviews (Bailey, 1971; Jaques and Morris, 1981; Kaya, 1982; Flexner *et al.*, 1986; Harper, 1986; Laird *et al.*, 1990; Brooks, 1993; Rosenheim *et al.*, 1995; Begon *et al.*, 1999). With justifiable emphasis on environmental safety, a great deal has been written on the real and potential negative interactions between these groups of natural enemies. The objective of this chapter is to present an overview of the diverse positive interactions between pathogens and entomophagous insects, meaning by "positive" those interactions with known or potential capacity to enhance biological control. Also addressed are a number of inconsistencies in the applications of the terminology of interaction and in methods of analysis. No attempt has been made to present a comprehensive review of the extensive and growing body of literature on pathogen–parasite–predator interactions or to balance the potential positive with the potential negative aspects of any specific natural enemy combination. All such assessments must ultimately be left to the research arena.

It should be noted that this work focuses on interactions between pathogenic microorganisms and parasitic or predatory insects, and thus excludes interactions between nematodes (generally studied by insect pathologists) and entomophagous insects. A few natural-enemy associations used as examples of specific types of interactions include nematodes; however, in these, the nematode is viewed as a parasite rather than as a pathogen.

Terminology

Clear communication among scientists investigating interaction phenomena is essential and would obviously be facilitated by standardization of terminology. Such standardization has never been achieved, leading to much confusion with regard to analysis, interpretation, and communication of experimental results. It is not my intent in this short review to attempt a resolution of this difficult problem; however, it is necessary to define the key terms that will be used in this paper.

Synergistic effects

Two or more biological control agents can be said to act synergistically in the regulation of a pest population when the control effects they exert in combination exceed the sum of the effects that would be expected were they acting completely independently (exhibiting independent, uncorrelated joint action) (see Robertson and Priesler, 1992). In simplest terms (and in the familiar parlance of ANOVA (analysis of variation)), the effects caused by one agent are dependent upon the level of another agent. When the net result of the dependent actions is enhancement of an effect, the result is called synergism. If the dependent actions result in inhibition of the effect, it is described as interference or antagonism (Sokal and Rohlf, 1995). Effects may be expressed in various ways, including mortality or reduced time to death, or even in terms of crop damage or yield losses. Synergism in which the result of dependent action was more rapid expression of an effect than would occur were each agent acting independently has been called temporal synergism (Benz, 1971). Synergism is normally discussed in terms of two control agents and the interdependence of their effects; however, synergism between multiple control factors is certainly possible, especially in field environments with multiple interacting agents.

A synergistic effect can sometimes result from combined actions of two or more agents even when their interaction does not directly affect the susceptibility of a host or the infectious, parasitic, or predatory capacity of a control agent. Interactions in which species do not physically interact have been termed indirect interactions (Wootton,

1994). Perhaps synergism resulting from such interactions should be classified as indirect rather than true synergism. However, no distinction will be made here when the result is clearly synergism in terms of the ultimate level of pest population or crop damage reduction and satisfies the definition of synergism in the ANOVA context: whenever the effect of one agent depends (in any way) on the level of another.

Economic synergism is a useful term applied to synergistic effects observed in field or greenhouse environments (see Benz, 1971). This refers to synergism measured not by mortality or detectable reductions in populations of a specific pest, but rather in terms of crop damage or yield increases; that is, a combination of control agents produces greater crop protection than would be achieved by each agent operating independently. The utility of this term is readily apparent in field research, where the various control agents and the nature of all the specific interactions producing a synergistic effect may be largely unknown.

Other beneficial combined effects

No standard terminology has been established to refer to subsynergistic interactions among natural enemies, which are probably far more prevalent than synergism and much more important in terms of biological control. Many of the terms commonly used to describe interactions between pest control agents were introduced many years ago by researchers in the fields of pharmacology and toxicology. In his studies of the toxic effects of chemical mixtures, Bliss (1939) defined three principal types of joint action:

- Independent joint action
- Similar joint action
- Synergistic action

The concepts of independent joint action and synergistic action were presented above. Similar joint action relates to mixtures of two materials containing active ingredients with similar modes of action. In such cases, the observed response is essentially the same as would result if the organism were exposed to an amount of active ingredient equal to the sum of the quantities present in each material. This later came to be called an additive effect (Robertson and Priesler, 1992; Greco *et al.*, 1992), and this definition of the term has persisted (Pöch *et al.*, 1995).

Entomologists and ecologists have faced a difficult dilemma in adopting the inter-action concepts and definitions developed by toxicologists. Interaction analysis is based upon rejection of a hypothesis of independent action, and a greater than hypothesized effect is called synergism. But what should one call an effect that is less than predicted by the hypothesis? Levels of control less than the combined independent effects must technically be considered antagonism (Robertson and Priesler, 1992; Sokal and Rohlf, 1995). However, the term antagonism is not very useful in the context of pest control, as it applies to any observed level of control from zero to the level of the combined independent probabilities! Moreover, the term antagonism seems inappropriate to describe combined effects that may be highly beneficial even if not achieving the level of synergism. It is not surprising, therefore, that in studies of interactions among biological control agents, entomologists have most commonly

Table 8.1 Definitions of terms used in reference to control agent interactions

Interaction	Pest control effect observed in a two-agent system
Synergistic	Effect greater than the level predicted by the model of independent, uncorrelated joint action (combination of independent probabilities)
Independent additive	Effect equal to the level predicted by the model. Equals Bliss independence of Greco *et al.* (1992)
Additive	Effect less than predicted by the model and greater than that of the more effective agent alone[1]
Nonadditive (neutral)	Effect from two agents combined is unchanged from the effect of the more effective agent alone
Negative	Effect from two agents combined is less than the effect of the more effective agent alone.

[1] Considering Greco *et al.* (1992), this might be termed Bliss additivity.

referred to the effects produced by agents exhibiting independent joint action as additive effects. This term is also applied quite broadly to effects that are less than predicted by independent action but are still beneficial (resulting from incomplete or low levels of antagonism). Such effects also have been called subadditive. Use of the term additive in this way is obviously in conflict with the definition of additivity presented above. In the field of biological control, different organisms or microorganisms (and possibly even different strains of the same microorganism) used in combination usually have very different modes of action. Thus, accepting the pharmacological definition, the term additive would rarely be applicable to studies of complex interactions between biological control agents.

After consideration of this problem, it was nevertheless considered unnecessary to attempt introduction of any new terminology. Use of the term additive is too well established in the literature to be discarded, and it is also simply the most efficient and descriptive term for relating the concept of beneficial combined effects. However, even this term is interpreted differently by different entomological and ecological researchers, and for clarity, a definition of terms used herein is presented in Table 8.1.

Analysis of synergism

Two or more lethal agents exhibiting uncorrelated joint action produce a combined effect on a host population defined by the combination of independent probabilities (Robertson and Priesler, 1992). In many analyses, this hypothesized level of combined effects represents the critical threshold value for assessing synergism. It is important to keep in mind that the individual probabilities are only theoretical or potential rates determined under controlled experimental conditions in the laboratory or in manipulative field experiments (holding the host and each control agent together in isolation). It also is important to remember that combination of independent probabilities predicts a combined rate that is less than the simple arithmetic sum of the probabilities (see equation (1) below). When the hosts are exposed to the combined actions of the natural enemies, the model predicts that a certain percentage will, by chance alone, be attacked by more than one natural enemy. Thus, for example, a level of pest control predicted by this model may occur in a system in which many parasitized hosts are lethally infected by a pathogen (resulting also in the death of the parasite). This is counterintuitive, but exactly the outcome expected when one tests a hypothesis of independent, uncorrelated joint action (a hypothesis of no interaction)! In the absence of rigorous analysis, one must, therefore, resist concluding that observed competition among natural enemies is indicative of an overall antagonistic relationship.

These theoretical rates (independent probabilities) can be determined with reasonable accuracy when the measured effect is mortality among members of a single pest cohort. The difficulty of estimating such rates is evident, however, when the desired effect is control of a multivoltine pest population or control of damage inflicted on a crop. Biological systems are, moreover, inherently complex, with numerous species interacting within and across trophic levels. In studies of such systems, it is usually impractical or impossible to test all of the key combinations of species. Thus, while evidence of synergism may be found in a biological control system, its statistical proof may remain elusive. Even more significantly, synergistic interactions identified in laboratory studies may not exist (or exist consistently) under field conditions. Path analysis, a powerful analytical method based on multiple regression, can aid in the prediction and elucidation of key interactions in complex field systems (Wootton, 1994; Sokal and Rohlf, 1995).

Synergism between two control agents is most commonly detected and evaluated by biological control researchers using two well-known statistical procedures, chi-square or G analysis of frequencies (essentially goodness-of-fit tests) and two-way ANOVA with analysis of interaction (Sokal and Rohlf, 1995). The procedures test hypotheses of either no interaction (ANOVA) or independent action (chi-square or G) of the control agents. Rejection of the hypothesis indicates that the actions of the two agents are interdependent.

Use of ANOVA for detection of synergism usually involves a straightforward multiway design with the factors (e.g., control agents A and B) divided into classes (or levels). In simplest form, simply the presence or absence of each factor defines treatments. Treatments corresponding to presence of each agent alone and in combination and a control including neither agent are required to complete a 2×2 factorial design. Replication in field experiments is typically generated through blocking (randomized complete block design). The test of synergism involves a standard F test of the hypothesis of no interaction ($A \times B$).

Chi-square analysis is somewhat more complicated but nevertheless relatively simple to conduct. The theory of independent probabilities is applied to calculate the expected level of effect (e.g., proportional mortality) from all mortality factors acting jointly but independently. Effects of natural (control) mortality factors must also be included (Robertson and Preisler, 1992).

$$P_{expected} = P_C + P_A(1 - P_C) + P_B(1 - P_C)(1 - P_A) \qquad (1)$$

where P_C = proportion of insects affected in the controls; P_A = corrected proportion of insects affected by control agent A; and P_B = corrected proportion of insects affected by control agent B.

Failure to detect some significant synergistic interactions may result unless the rates of mortality observed in each treatment group are corrected for control mortality. This is accomplished using the familiar Abbott correction, which is simply an application of the standard rule for combination of independent probabilities (Finney, 1971). Use of uncorrected proportions inflates the expected proportion and thus ultimately affects the calculated chi-square value and the test of the hypothesis of independent action.

The chi-square statistic is calculated using the standard formula incorporating observed and expected frequencies (see Sokal and Rohlf, 1995).

$$\chi^2_{[1]} = \frac{(L_o - L_e)^2}{L_e} + \frac{(D_o - D_e)^2}{D_e} \tag{2}$$

where L_o = observed corrected number of unaffected insects; L_e = expected corrected number of unaffected insects; D_o = observed corrected number of affected insects; and D_e = expected corrected number of affected insects.

There has long been a tendency by researchers using this common method of synergism analysis to indiscriminately pool data from replicated treatments and, in some cases, to apply the analysis to unreplicated treatments. Such approaches may represent serious forms of pseudoreplication (Hurlbert, 1984). An analytical method employing the G-test statistic for replicated tests of goodness of fit is described by Sokal and Rohlf (1995). Another common error is the failure to include both contributions to chi-square from the alternative responses (in this example, affected and unaffected insects).

Synergism, as defined above, may result from a broad range of interactions among the various components of a biologically based pest management system. Those with greatest importance to biological control are described below. As indicated in the introduction, this discussion deals with both known and hypothetical interactions. Many of the described interactions have not been proven to result in synergism in terms of pest or crop damage control.

Synergism resulting from interactions between natural enemies and the host

Interactions affecting host physiology/immunity

Synergism between two or more control agents most commonly occurs as the result of some direct joint action against the target pest. This is the classic mode of action of chemical insecticide synergists. The synergist, which may be nontoxic, disrupts the detoxification processes of the insect, making it highly susceptible to a jointly applied chemical insecticide. The closest analogy in a pathogen–entomophagous insect system would be a situation in which microbial infection compromises the immune system of an insect, leaving it more vulnerable to an invading parasite or, conversely, situations in which parasitism or predation leaves a host more susceptible to microbial infection. Toxins produced by insect pathogenic bacteria or fungi are known to inhibit cellular immune responses in insects (Boucias and Pendland, 1998), and immunologically suppressed hosts are more vulnerable to a broad range of antagonists. The potential significance of such interactions is exemplified by the remarkable obligate mutualistic relationship between many endoparasites and polyadnaviruses. These viruses block the host cellular immune response, allowing parasitoid eggs to evade encapsulation (Lavine and Beckage, 1995). However, to my knowledge, no such interaction classifiable as synergistic has been described between an insect microbial control agent and an insect parasite or predator. Endoparasites and their mutualistic symbionts are highly capable of overcoming host defense reactions without relying on other natural enemies to weaken the host, and it is usually disadvantageous for a parasite to select a host infected with a pathogen, as victory in such a competition is rarely certain. This may not be the case, however, with weak pathogens or pathogens that typically develop slowly in the infected host (or with pathogens whose development is inhibited owing to suboptimal environmental conditions). It has been hypothesized, for example, that *Bacillus popilliae* infection weakens the midgut wall of scarab grubs, making them more susceptible to penetration by parasitic nematodes (Thurston *et al.*, 1994). White grubs stressed by exposure to the facultative pathogen *Bacillus thuringiensis japonensis* have been shown to be highly susceptible to nematodes owing to unknown mechanisms (Koppenhöfer and Kaya, 1997). Such effects of pathogenic microbes or microbial toxins on the wall of the gut or other body parts may also predispose a host to invasion by parasitoids. Gösswald (1934), for example, reported that *Sarcophaga shutzei* invaded only virus-infected hosts.

Potential for synergistic interaction also exists in the opposite situation, in which the effects of a predator or parasite increase host susceptibility to a pathogen. Any injury that breaches the body wall obviously predisposes a host to microbial infection. Injuries are commonly inflicted by the actions of invading parasites. Perforation of the gut wall by tachinid larvae has been shown to predispose larval gypsy moth to infection by bacteria or nucleopolyhedroviruses (Godwin and Shields, 1984). King and Bell (1978) demonstrated that larval *Heliothis zea* (Boddie) parasitized by the braconid parasite *Microplitis croceipes* (Cresson) were more susceptible to fungal infection than were healthy larvae, but the mechanism underlying this phenomenon was not elucidated. A series of studies (Führer and El-Sufty, 1979; El-Sufty and Führer, 1981, 1985) determined that the integuments of parasitized lepidopteran larvae were more easily penetrated by the fungus *Beauveria bassiana* as a result of various mechanisms; however, the fungus was subsequently inhibited by a fungistatic factor produced by the parasitoid (see below).

The direct inoculation of a pathogen into the body of an insect via the ovipositor of a parasite contaminated or infected with a pathogen represents an interaction that is potentially synergistic, as the first line of defense of the insect body (the integument) is breached in dramatic fashion. There exists an extensive body of literature on this phenomenon, especially pertaining to microsporidia and viruses (see Brooks, 1993; Tanada and Kaya, 1993). Much of the advantage gained by the pathogen from such interactions can be attributed to increased dispersal, an aspect covered in a later section under interactions between natural enemies.

Interaction affecting host development

Interference with normal host development is a common result of parasitism or infection. Insects so affected may be more susceptible to a variety of natural enemies, especially those that target the larval stages of their prey. In most cases, insects with microbial infections so severe as to inhibit development ultimately succumb to the infection; thus, the interaction between microbe and predator or parasite rarely becomes synergistic in terms of ultimate reductions in pest numbers. On the other hand, temporal synergism might be detectable for at least a brief period if infected insects are more readily preyed upon than uninfected insects during the period of disease incubation. Weiser and Veber (1957) observed that fall webworm larvae infected with a microsporidian were more susceptible to predation, although it was suggested that the consequent reduction in the reproductive potential of the protozoan was a possible disadvantage. The same effect could also result when parasitized insects become infected with a pathogen. Hochberg (1991a), for example, determined that the lifespan of *Pieris brassicae* (Linnaeus) larvae harboring both the parasite *Apanteles glomeratus* (Linnaeus) and a granulovirus was shorter than that of larvae infected only by the virus.

Synergism appears relatively common between entomophagous insects and microbial control agents whose primary mode of action involves toxinosis. Sublethal doses of toxins often result in prolongation of development. Insects so affected may be more susceptible to attack by predators and parasites. Weseloh and Andreadis (1982) and Weseloh *et al.* (1983) observed that *Bacillus thuringiensis* intoxication retarded development of gypsy moth larvae. They concluded that a synergistic interaction observed between the pathogen and the parasite *Apanteles melanoscelus* (Ratzeburg) resulted from the extended development of the larvae and thus longer exposure to attack by the parasite. Similar interactions were observed with *B. thuringiensis*, the parasitoid *Rogas lymantriae*, and gypsy moth (Wallner *et al.*, 1983) and with *B. thuringiensis*, the predator *Perillus bioculatus* (Fabricius), and Colorado potato beetle larvae (Cloutier and Jean, 1998). Potential for a similar interaction between *B. thuringiensis* and the tachinid parasitoid *Myiopharus doryphorae* (Riley) was suggested by Lopez and Ferro (1995). Since insects continue to feed and cause crop damage after becoming infected with most pathogens, the rapid elimination of infected insects by predators during this period could also result in important levels of economic synergism.

As explained above, the model of independent, uncorrelated joint action between two or more natural enemies actually predicts a probability of mortality of the targeted host that is less than the arithmetic sum of the independent probabilities (equation (1)). Independent joint action is possible only under the condition that each control agent does not act in any way (directly or indirectly) that affects another

S.P. Wraight

agent's selection or utilization of a host. Any mechanism, therefore, that minimizes contact between mutually antagonistic control agents has potential to produce a synergistic interaction. Insects infected with pathogens may exhibit various symptoms that leave them undesirable to parasites or predators. One such symptom is reduced host size. Stark *et al.* (1999) observed that larvae of the western grape skeletonizer, *Harrisina brillians* (Barnes & McDunnough) infected with Hb-granulovirus (HbGV) were smaller than healthy larvae and were avoided by a tachinid parasitoid, *Ametadoria misella*, known to prefer older (larger) larvae. It was concluded that this phenomenon reduced overlap of the parasitoid and HbGV and allowed the two control agents to coexist and significantly impact pest populations. Stunted survivors of bacterial toxinosis also may be heavily targeted by parasites that prefer early instar (small) hosts (Weseloh and Andreadis, 1982).

Potential also exists for synergistic interactions in the alternative situation, in which hosts affected by parasites become more susceptible (or remain susceptible) to other natural enemies. Host manipulation by a parasite may induce the host to remain longer in the larval feeding stage (Beckage, 1985). As related above, such larvae are potentially more susceptible to predators. Additionally, however, these effects may increase their susceptibility or vulnerability to a broad range of pathogens. If uninhibited by the parasite, an invading pathogen could quickly kill the host, resulting in temporal synergism.

Effects of host manipulation may be manifested as protracted intermolt periods. Host molting is a well known mechanism whereby insects resist fungal infection (Vey and Fargues, 1977; Vandenberg *et al.*, 1998), and less frequent molts may increase an insect's susceptibility to these pathogens. To my knowledge, synergism between a parasite and pathogen has never been attributed to this mechanism; however, synergism observed between entomopathogenic bacteria and fungi (Glare, 1994; S.P. Wraight, unpublished) may be due, at least in part, to such an effect.

Interactions affecting host behavior

The debilitating effects of microbial infection or intoxication described above also commonly include abnormal behavior and, clearly, a great range of altered behaviors resulting from microbial infection or toxinosis have considerable potential to increase insect susceptibility to various natural enemies. Lopez and Ferro (1995) and Cloutier and Jean (1998) demonstrated that weakened, *B. thuringiensis*-intoxicated Colorado potato beetle larvae were less able to physically repel attacks by a tachinid parasitoid and a hemipteran predator, respectively, than were healthy larvae. As described above, significant mortality caused by other control agents before the insects would normally succumb to infection might result in economic synergism.

In addition to the potential synergism resulting from reduced activity, synergistic interactions may also derive from excited behavior (escape responses) of the host induced by natural enemies. From results of laboratory studies, Furlong and Pell (1996) concluded that the presence of a foraging parasitoid, *Diadegma semiclausum* Hellen, increased movement of diamondback moth larvae and thus infection by the fungus *Zoophthora radicans* distributed contagiously in the test environment. Roy *et al.* (1998) suggested a similar mechanism for increased infection of pea aphids exposed to the predator *Coccinella septempunctata* Linnaeus in the presence of the fungus *Erynia neoaphidis*. In such experiments, however, it is difficult to distinguish effects resulting from increased movement from effects caused by mechanical transmission of the pathogen by the predator.

Orthopterans and other insects infected with pathogens exhibit a distinct behavior commonly referred to as behavioral fever. Infected individuals climb to the tops of vegetation to bask in sunlight (Bronstein and Conner, 1984; Bronstein and Ewald, 1987; Carruthers *et al.*, 1992; Inglis *et al.*, 1996). Such behavior can raise internal body temperatures to levels that are inhibitory to invading pathogens. Infections by heat-sensitive pathogens may even be cured in this way. Thus, any agent capable of interfering with or preventing the normal basking behavior of insects infected with heat-sensitive microbes could function as a synergist. Basking, at the same time, exposes insects to greater risk of predation or parasitism (Carruthers *et al.*, 1992).

Many parasitoids are capable of inducing even more pronounced and abnormal behavioral changes in their hosts, including stimulation of the host to seek concealed locations protected from adverse weather conditions (Feener and Brown, 1997). This behavior could expose insect to pathogens (possibly to pathogen reservoirs) under conditions highly favorable to infection or disease development.

Synergism resulting from actions affecting host population dynamics

Parasitic and predatory insect populations generally cycle in a host density-dependent manner, affecting a greater percentage of hosts as the host population increases. This negative feedback phenomenon generates powerful regulatory pressures on insect populations, in many cases preventing them from reaching economically damaging levels (Berryman, 1999). This is particularly true for highly efficient hunters whose density-dependent actions may be effective over a range of very low host densities (DeBach and Rosen, 1991). At the same time, however, many natural enemies have lower population growth potential than their hosts (owing to lesser fecundity or longer developmental cycles), and their populations characteristically lag behind host populations. Under certain situations, especially in fields of short-cycle crops and in regions with short growing seasons, pest populations may "escape" control by such lagging enemy populations and quickly reach severely damaging outbreak levels.

This is a common problem associated with biological control, and one of the avenues pursued as a solution has involved introductions or augmentative releases of multiple control agents. In such programs, there is potential for a broad range of interactions among these agents, including synergism. Perhaps the most easily envisioned mechanism for synergism under these circumstances is one in which an agent of moderate efficacy (such as a predator with low fecundity) acts to suppress the growth of a pest population, allowing the population of a lagging but ultimately more efficient agent to reach an effective level before the pest is able to escape control. Another example would be establishment or periodic release of two control agents that are each effective during different parts of the growing season (e.g., when environmental conditions are most favorable to their activity). Acting alone, each of these agents might be unable to prevent a pest from escaping, whereas, acting jointly, they are capable of effecting long-term control. Begon *et al.* (1996) found that a baculovirus and parasitoid attacking different life stages of a host could drive laboratory populations to extinction.

Natural enemies operating during different seasons, or in different habitats (e.g., one agent controlling the adult stage of a pest in the crop canopy and another controlling immature stages in the soil) are often referred to as complementary rather than synergistic. It seems reasonable, however, to consider such complementary relationships as being at least indirectly interactive. Such effects of one agent on a host population may have very definite indirect effects on the biological control efficacy of a control agent that acts at a later time in a host population cycle or life cycle. Season-long pest control or crop damage control resulting from independent, essentially separate (nonjoint) action of control agents is clearly a difficult issue with respect to analysis of biological interactions. Evaluation of the significance of enhanced effects from complementary and other indirect effects would seem to call for a different null hypothesis than that represented by the model of independent joint action (equation (1)). Complementarity is widely recognized as a highly beneficial phenomenon that enables coexistence of natural enemies (Begon *et al.*, 1999). It should be noted, however, that the term complementarity, like additivity, is not used consistently by researchers, and is also applied more broadly to associations in which natural enemies are isolated by various intrinsic and extrinsic mechanisms.

Various pathogenic microbes have considerable potential to suppress pest populations and, as mentioned, allow slow-developing populations of entomophagous insects to become effective. The effectiveness of these agents, like that of parasites and predators, is directly related to host density. However, because many pathogens cycle more rapidly than other natural enemies and are capable of producing great numbers of infectious units that can effectively saturate a host's environment, pathogens have potential to regulate even those pest populations that have escaped control by other agents. Microbial control of such populations typically occurs too late to protect a crop from damage; however, benefits may be realized from reductions in future pest populations.

Many microbes have the advantage over parasites and predators of potentially explosive population growth and thus the capacity to rapidly control very high-density host populations. They are disadvantaged in that they are destroyed or inhibited by environmental conditions normally prevalent in many agroecosystems (e.g., high solar irradiance, high temperature, low relative humidity) and are also unable to actively search for hosts. Consequently, few pathogens are capable of operating as effective control agents at very low host densities. This difference in the modes of action of pathogens versus entomophagous insects provides opportunity for synergistic interaction even beyond the examples mentioned above. For example, spray applications of pathogens might successfully control pests residing in exposed parts of the crop canopy and force parasites or predators to seek out those in more cryptic habitats that might otherwise remain unaffected. Entomophagous insects may also vector pathogens into these habitats (see below). The result could be a greater total effect on the pest population than would be expected from the combined independent action of the control agents. In his review of the epizootiology of insect diseases, Tanada (1964) cited observations by Niklas (1939) and Franz and Krieg (1957) that tachinid parasitoids seeking hosts moved out of areas with virus epizootics. A complex of parasites and predators forced to compete for hosts whose numbers have been substantially and suddenly reduced as the result of a disease epizootic (e.g., resulting from a microbial control application) have the potential to essentially eradicate a local pest population and provide control for a prolonged period. Such effects, of course, also place populations of other natural enemies at risk.

Synergism resulting from interactions between natural enemies

Interactions affecting natural enemy dispersal

Host–parasitoid–pathogen models have indicated that coexistence of parasitoids and pathogens is enhanced when both natural enemies have high search/transmission and survival rates (Hochberg *et al.*, 1990). With respect to insect pathogens, both transmission and survival may be increased through various associations with natural enemies. Such effects result from the capacity of predators and parasites to transport pathogens. The infectious units of various microbial control agents, including phoretic nematodes, virus occlusion bodies, and bacterial, protozoan, and fungal spores, may simply be picked up and transported on the surface of the insect's body. Also, many pathogen propagules survive passage through the digestive tracts of entomophagous insects and are deposited in the feces. In the case of pathogens with broad host spectra, the insect vector may itself become infected, succumb to infection, and support production of infectious propagules. In this case, greater numbers of infectious units may be released into the environment (but at the expense of the parasite synergist).

As discussed earlier, the oviposition and feeding activities (host wounding) of parasites have been extensively hypothesized as mechanisms of pathogen transmission, especially of microsporidia (see Brooks, 1993). The results of studies with many pathogens are inconclusive, however, and in terms of epizootiology this mechanism may be no more important than simple mechanical transmission of pathogens onto the food or into the microhabitats of insects (see Entwistle and Evans, 1985; Hochberg, 1991b; Brooks, 1993). In addition to promoting dispersal, transport or release of propagules into a protected environment has potential to affect the persistence of the pathogen. Natural enemies must also be included among the important biotic agents identified by Hochberg (1989) as aiding movement of pathogens into and out of important reservoirs.

The phenomenon of pathogen vectoring by natural enemies is given considerable emphasis in most reviews of epizootiology of insect diseases (Tanada, 1964; Andreadis, 1987; Tanada and Kaya, 1993). It has been highlighted as the most prominent positive interaction among pathogens and other natural enemies in recent reviews of pathogen–predator–parasite interactions (Brooks, 1993; Roy and Pell, 2000). The vectoring actions of entomophagous insects have considerable potential to introduce pathogens into healthy pest populations and promote development of disease epizootics. Releases of ichneumonid and braconid parasites contaminated with nucleopoly-hedroviruses (NPVs) resulted in successful transmission of the pathogens into pest populations in the field (Mohamed *et al.*, 1981; Young and Yearian, 1990). Biever *et al.* (1982) initiated an epizootic in a cabbage looper population by releasing pentatomid predators, *Podisus maculiventris* (Say), contaminated with cabbage looper NPV. Roy *et al.* (2001) demonstrated that adult coccinellid beetles could vector the fungus *Erynia neoaphidis* to aphids in field cages. Pathogen vectoring via the host-searching activities of natural enemies has potential to initiate epizootics in low-density host populations at times earlier than those at which they would otherwise be initiated and in cryptic habitats that the pathogen would normally be unable or slow to invade. Such interactions clearly have considerable potential to produce synergistic results.

Interactions affecting natural enemy behavior or development

As stated at several points previously, mechanisms that prevent a single host from being attacked by more than one species of natural enemy prevent potentially negative interactions between the agents and thus may be highly beneficial in terms of biological control. Such mechanisms have been described as antiantagonistic (Mesquita and Lacey, 2001). The model of independent joint action predicts the chance selection of some individuals by more than one agent, and potential for synergism exists whenever one natural enemy has the capacity to avoid a host already under attack by another agent and seek another, healthy host.

Numerous studies have documented the ability of various parasitoids to identify and avoid or reduce oviposition or larviposition in hosts infected with diverse pathogens, including baculoviruses (Kelsey, 1962; Versoi and Yendol, 1982; Caballero *et al.*, 1991; Sait *et al.*, 1996), bacteria (Temerak, 1980, 1982; Salama and Zaki, 1983; Lopez and Ferro, 1995), protozoa (Laigo and Paschke, 1968), and fungi (Brobyn *et al.*, 1988; Fransen and van Lenteren, 1993; Mesquita and Lacey, 2001). In many of these cases, however, discrimination was evident only when parasites were presented with hosts in advanced stages of infection. Thus, these mechanisms are, at best, only partially effective in preventing antagonistic interactions.

Because predators do not face the same risks as parasitoids in attacking a host infected with a pathogen, their capacity to differentiate between infected and healthy hosts has received less attention. Predators will generally feed on hosts with both early and advanced stages of infection (Abbas and Boucias, 1984; Gröner, 1990; Roy and Pell, 2000). However, predators may reject hosts that have succumbed to infection. The lacewing *Chrysoperla carnea* (Stephens), syrphid fly *Episyrphus balteatus* (DeGeer), and ladybird *Hippodamia convergens* (Guérin-Méneville) did not feed, and *Coccinella septempunctata* (Linnaeus) fed only to a limited extent, on aphid cadavers with sporulating fungal pathogens (Roy *et al.*, 1998; Pell and Vandenberg, 2002).

Some parasites are also known to produce chemical substances in the parasitized host that inhibit development of competing pathogens. Many parasitic hymenoptera, for example, release specialized cells called teratocytes into the host, which produce fungistatic and possibly bacteriastatic compounds (Dahlman and Vinson, 1993). An anal secretion from larvae of the Ichneumonid *Pimpla turionella* (Linnaeus) also exhibits antimicrobial properties (Willers *et al.*, 1982; Führer and Willers, 1986). These substances are probably responsible for observations of parasitized hosts becoming highly resistant to fungal infection (Fransen and van Lenteren, 1994; Jones and Poprawski, 1996). Results of studies with viruses suggest that parasites may produce antiviral factors or alter host physiology in ways that inhibit virus development (Beegle and Oatman, 1974; Teakle *et al.*, 1985; Santiago-Alvarez and Caballero, 1990). These interactions cannot be classified as synergism, because the efficacy of the pathogen is not enhanced (a pathogen prevented from successfully infecting a parasitized host cannot leave the host and search for another). On the other hand, this interaction is obviously important in preventing the competitive displacement of the parasitoid by the pathogen (especially during disease epizootics), and contributes balance in the pathogen–parasite association and greater potential for coexistence (Hochberg *et al.*, 1990; Begon *et al.*, 1999). Also, synergism resulting from long-term effects on parasite population dynamics cannot be ruled out.

Mechanisms that permit an entomophagous insect and a pathogen to successfully utilize the same host are similarly beneficial. Many tachinid parasites are able to develop in heavily infected or even pathogen-killed hosts, probably owing to their capacity to obtain air from outside the host (Entwistle, 1982). Accelerated development of parasitoid larvae in virus-infected hosts was recently observed by Escribano *et al.* (2000), and these researchers proposed that this might represent an additional survival strategy for parasitoids developing in infected hosts. Entwistle (1982) suggested that parasite preference for young hosts, e.g., by *Apanteles* spp., evolved as a mechanism to avoid virus-infected hosts. He referred to these parasites as "isolationists" and contrasted them to "nonisolationists" such as tachinid parasites capable of tolerating infections.

Interactions affecting higher-level predators or parasites (hyperparasites)

Under natural conditions, insect-pest populations are usually regulated by a broad range of predatory, parasitic, and pathogenic organisms; these, in turn, may be regulated by antagonists of a higher trophic level. In this situation, control agents with broad host ranges may interfere with these complex interactions. A chemical or biological control agent, could, for example, through a substantial independent lethal effect on an important hyperparasite, indirectly enhance the efficacy of a second control agent that is normally suppressed by the hyperparasite. While such trophic cascades result from effects that are independent at the organismal level, the actions of the two agents are clearly interdependent. Enhanced control results because the first agent, by removing the hyperparasite, effectively synergizes the predatory activities of the second agent. Synergism, in this case, can be considered the result of indirect interactions (Wootton, 1994). Such positive effects have been reported as a result of interactions between entomophagous insects, but I am not aware of any examples involving pathogens.

Interactions affecting intraguild predators and parasites

The above discussion of interactions between organisms at different trophic levels generally applies also to organisms comprising a natural enemy guild. By preying upon or infecting an important intraguild predator (in addition to the target pest), a control agent has potential to indirectly enhance the efficacy of a second control agent that is normally suppressed by the intraguild predator. Such interactions are most likely to occur among generalist agents with very broad host ranges and, most likely, as an unexpected consequence of a biological control program. Such effects might be revealed as more complex analyses of systems receiving inundative applications of microbial pesticides are conducted.

Other beneficial interactions

The degree to which interactions between natural enemies actually fit the independent joint action model under natural conditions is difficult to assess and largely unknown. Independent joint action may be rare ("it's a jungle out there," and antagonism rules) or it may be more common than widely believed (the model predicts a certain level of random interaction or competition between control agents). Most would agree, however, that antagonistic interactions among biological control agents are very common (more common than synergististic effects), and represent one of the most important constraints to biological control. Interference among biological control agents in the form of intraguild parasitism, predation, or infection is so commonly encountered that many researchers have recommended development only of specialist agents, or use of generalists only as a second line of defense against a pest that has escaped control by a specialist. Many natural enemies introduced or applied for microbial control (e.g., many hyphomycete fungal pathogens or coleopteran predators) have rather broad host ranges, and may be important intraguild predators under certain condition.

Considering the predominance of negative interactions among biological control agents, independent additive effects, and a broad range of additive effects (Table 8.1) are clearly highly desirable outcomes. In biological control systems, the levels of control provided by two agents capable of operating more or less independently may be as valuable as the enhanced levels achieved through a synergistic interaction. In fact, any additive effects may be more valuable than synergism if the association producing these effects is more stable and able to provide long-term control without need of intervention.

Combination of pathogens and entomophagous insects for biological control

This is clearly a topic that cannot be fully developed in this brief account of beneficial interactions between pathogens and other natural enemies. Excellent discussions are presented by Hochberg (1989), Hochberg *et al.* (1990), Rosenheim *et al.* (1995), Begon *et al.* (1999), and Roy and Pell (2000).

In biological systems, virtually all species are under pressure from multiple natural enemies and natural competitors. The diversity of interactions among these species is enormous and extraordinarily complex. Studies of natural enemy complexes have repeatedly revealed that the relative abundance and impact of each species changes between and even within seasons in response to interacting biotic and abiotic factors, many of which remain poorly understood. It thus seems highly likely that synergistic or antagonistic interactions between two or more natural enemies characterized in controlled experiments will prove as variable and unpredictable as the systems within which they occur and may appear and disappear, or become more or less intense, in response to such factors as weather, host density, natural enemy density, and host plant quality. Long-established interactions may break down when introduced or invasive species enter a community. From this follows the caveat that the efficacy and stability of specific natural enemy complexes may never be completely predictable.

It has been concluded that parasitoids are disadvantaged when competing with microbial control agents because they are highly vulnerable to premature host death (Vinson, 1990; Brooks, 1993). Yet there are many examples of parasitoids and pathogens operating jointly to provide effective biological control (see reviews by Tanada, 1964; Flexner *et al.*, 1986; Laird *et al.*, 1990). The potential for compatibility between pathogens and parasites has been supported by studies of theoretical models (Hochberg *et al.*, 1990; Begon *et al.*, 1999). According to Hochberg *et al.* (1990):

> Although pathogens often have the upper hand over parasitoids to multiply in infected hosts, our model suggests that it is a misconception to conclude that parasitoids will be eliminated from the system or play a minor role in the regulation of the host. Furthermore, the commonly held notion that natural enemies must not interfere with each other may be misleading; some interference, in terms of overlap of attack times and/or the possibility of survival in a co-infected host, can be a desirable characteristic of systems with multiple natural enemies.

These researchers went on to conclude:

> Biological control involving parasitoids and pathogens is most likely to be a sound strategy when
> (i)   the pathogen produces external stages that may span the intergenerational gap;
> (ii)   parasitoid (and pathogen) attacks are moderately clumped;
> (iii)   both natural enemies have high rates of search or transmission; and

(iv) there is some degree of overlap in the timing of attacks of the enemies, and/or competition with the host is not invariably one-sided.

While identifying the potential for pathogen–parasite coexistence, this and other models clearly indicate that addition of any pathogens to an existing biological control system must be effected with great caution to avoid disruption of an already functioning natural enemy complex (Hochberg *et al.*, 1990; Briggs, 1993; Hochberg, 1996; Begon *et al.*, 1997; Begon *et al.*, 1999; Onstad and Kornkven, 1999). Any attempt to exploit laboratory-discovered synergism or other beneficial interactions among natural enemies certainly warrants equal caution and extensive preliminary study. The situation is clearly less problematic in protected crops, where a competitively displaced control agent can usually be reestablished with little difficulty.

The relative strengths and weaknesses of pathogens, parasites, and predators in terms of various traits such as reproductive potential, speed of insecticidal action, and potential to function as synergists are presented in Table 8.2. These characteristics suggest some additional considerations with respect to the combination of pathogens and entomophagous insects for biological control.

Table 8.2 Relative strengths and weaknesses of parasites and predators compared to insect pathogens developed for microbial control

Control agent	Strengths	Weaknesses
Pathogens	• Potential for moderately rapid mode of action (toxin producers have rapid mode of action) • Potential for extremely high rate of reproduction • Generally strong intrinsic competitors • Many species easily and economically augmented • Generally easy to apply strategically • May be effective at low host density (via augmentation) • May be effective against pest outbreaks (via augmentation or epizootics) • May synergize activity of parasites and predators	• Generally unable to search for hosts (thus limited efficacy against cryptic pests) • Infectivity, speed of action, reproductive potential, effective field life, and thus biological control efficacy, generally dependent upon environmental conditions • May be important intraguild predators
Parasites	• Efficient searchers • In some cases, strong intrinsic competitors (produce antimicrobial factors) • Generally effective at low host density • May be able to detect and reject living hosts infected with pathogens • Not important intraguild predators • May synergize activity of pathogens	• Slow mode of action (slow development) • Low reproductive potential • In some cases, weak intrinsic competitors (susceptible to premature death of diseased host) • Generally not effective against pest outbreaks • Generally costly to augment
Predators	• Efficient searchers • Fast mode of action (predation) • May be effective at low host density • May be effective at high host density through augmentation • Some species may be applied strategically • May synergize action of pathogens	• Low reproductive potential • Generally costly to augment • May be important intraguild predators

It may be beneficial to select natural enemy combinations in which the pathogen functions as the synergist. Synergists, by enhancing the efficacy of a competitor, may tip the balance in favor of the competitor (and the more powerful the synergist, the greater the risk of displacement of the synergist). Considering the generally greater potential for augmentation of pathogens, it will usually be more practical and economical to replace a pathogen depleted by synergism (e.g., by spray application) than a parasite or predator. In fact, pathogens may deserve consideration as components of many biologically based pest control programs specifically because of the relative ease and precision with which their populations can be manipulated. A simple example of this approach would involve exploitation of the action of a pathogen that weakens the host, making it less able to resist parasites and predators. Rates and timing of microbial control applications that maximize the synergistic effect can be determined without concern for long-term survival of the pathogen.

Because parasites are often at a competitive disadvantage in hosts infected with pathogens, much research has focused on determining the time prior to infection needed by parasites to ensure their survival. The beneficial interactions reviewed here, however, suggest that it would likely be more effective to apply microbial control agents prior to the release or anticipated peak activity of entomophagous insects. If sufficient lead time were available to allow the pathogen to develop in the host, parasites or predators entering the system would be able to detect the pathogen and seek out the remaining healthy pests. Pests that inhabit the undersides of leaves and other cryptic habitats, are difficult to target with microbial control applications (Wraight and Carruthers, 1999), and these control agents might be highly complementary when used in this way. Fransen and van Lenteren (1994) developed such a strategy for combined use of the fungus *Aschesonia aleyrodis* and the parasite *Encarsia formosa* Gahan, for whitefly control in glasshouses. It was recommended that the fungus be applied 7 days prior to the parasite. Such a strategy may be incompatible with some crop production systems in which parasites alone can be relied upon for effective control of a key pest (e.g., van Lenteren *et al.*, 1996). Initial inundative releases of virulent pathogens could interfere with efforts aimed at early establishment of these parasites. In these systems, strategies calling for applications of pathogens followed by releases of predators might warrant consideration as a second line of defense against pest populations that have escaped control by the parasites or for control of secondary pest outbreaks. The relative ease with which microbial control agents can be mass-produced for augmentative releases allows for considerable strategic manipulation in a broad range of crop production systems. Applications may be timed to take advantage of known synergistic interactions. In the field, for example, a pathogen might be applied early in the season to preserve parasitoids that attack late instars of a pest (or vice versa).

Most parasites have long developmental cycles (and thus slow modes of action). Pathogens generally exhibit more rapid lethal action; however, most pathogens, including the fastest-acting (toxin producers) are most effective against the early instars of their hosts. Therefore, it is generally recommended, in the case of both pathogens and parasites, that augmentative releases be made during the early stages of a pest cycle. Thus, optimal integrated use of these agents will likely depend on discovery of pathogens and parasites capable of coexistence and joint action.

S.P. Wraight

Owing to such factors as the lethal effects of solar radiation, the weathering effects of wind and rain, the uneven distributions of their hosts, and the inefficiencies of microbial spray applications, pathogens typically occur in contagious distributions. This generally contributes to the heterogeneity of the system, a property known to promote coexistence of competing organisms (Atkinson and Shorrocks, 1981; Hanski, 1981; Ives and May, 1985). This fact, combined with the facts that (i) many larval parasites produce antimicrobial compounds in the host and (ii) adult female parasites are able to identify and reject hosts with advanced infections, indicates considerable potential for compatibility between selected pathogens and parasites (considering the Hochberg criteria presented above).

Conclusions

Pathogenic microorganisms and parasitic and predatory insects represent extremely different life forms that have developed a remarkable diversity of mechanisms and strategies to successfully compete for their insect hosts. A great array of negative interactions among insect pathogens and entomophagous insects has been extensively documented. However, from the detailed study of this competition in laboratory, greenhouse, and field arenas, biologists have identified an equally great variety of intrinsic and extrinsic interactions with positive and even synergistic effects on the capacity of these insect natural enemies to infect, parasitize, or prey upon a single host population. Study of these interactions is leading to development of novel strategies for successful integration of these control agents for improved biological pest control.

References

Abbas, M.S.T. and Boucias, D.G. (1984) Interaction between nuclear polyhedrosis virus-infected *Anticarsia gemmatalis* (Lepidoptera: Noctuidae) larvae and predator *Podisus maculiventris* (Say) (Hemiptera: Pentatomidae). *Environ. Entomol.*, **13**, 599–602.

Andreadis, T.G. (1987) Key factors: transmission. In J.R. Fuxa and Y. Tanada (eds.), *Epizootiology of Insect Diseases*, Wiley, New York, pp. 159–176.

Atkinson, W.D. and Shorrocks, B. (1981) Competition on a divided and ephemeral resource: a simulation model. *J. Anim. Ecol.*, **50**, 461–471.

Bailey, L. (1971) The safety of pest–insect pathogens for beneficial insects. In H.D. Burges and N.W. Hussey (eds.), *Microbial Control of Insects and Mites*, Academic Press, New York, pp. 491–505.

Beckage, N.E. (1985) Endocrine interactions between endoparasitic insects and their hosts. *Annu. Rev. Entomol.*, **30**, 371–413.

Beegle, C.C. and Oatman, E.R. (1974) Differential susceptibility of parasitized and nonparasitized larvae of *Trichoplusia ni* to a nuclear polyhedrosis virus. *J. Invertebr. Pathol.*, **24**, 188–195.

Begon, M., Sait, S.M. and Thompson, D.F. (1996) Predator–prey cycles with period shifts between two- and three-species systems. *Nature*, **381**, 311–315.

Begon, M., Sait, S.M. and Thompson, D.F. (1997) Two's company, three's a crowd: host–pathogen–parasitoid dynamics. In A.C. Gange and V.K. Brown (eds.), *Multitrophic Interactions in Terrestrial Systems*, Blackwell Scientific, Oxford, pp. 307–332.

Begon, M., Sait, S.M. and Thompson, D.F. (1999) Host–pathogen–parasitoid systems. In B.A. Hawkins and H.V. Cornell (eds.), *Theoretical Approaches to Biological Control*, Cambridge University Press, Cambridge, UK, pp. 327–348.

Benz, G. (1971) Synergism of micro-organisms and chemical insecticides. In H.D. Burgess and N.W. Hussey (eds.), *Microbial Control of Insects and Mites*, Academic Press, New York, pp. 327–355.

Berryman, A.A. (1999) The theoretical foundations of biological control. In B.A. Hawkins and H.V. Cornell (eds.), *Theoretical Approaches to Biological Control*, Cambridge University Press, Cambridge, UK, pp. 3–21.

Biever, K.D., Andrews, P.L. and Andrews, P.A. (1982) Use of a predator, *Podisus maculiventris*, to distribute virus and initiate epizootics. *J. Econ. Entomol.*, **75**, 150–152.

Bliss, C.I. (1939) The toxicity of poisons applied jointly. *Ann. Appl. Biol.*, **26**, 585–615.

Bronstein, S.M. and Conner, W.E. (1984) Endotoxin-induced behavioural fever in the Madagascar cockroach, *Gromphadorhina portentosa*. *J. Insect Physiol.*, **30**, 327–330.

Bronstein, S.M. and Ewald, P.W. (1987) Costs and benefits of behavioral fever in *Melanoplus sanguinipes* infected by *Nosema acridophagus*. *Physiol. Zool.*, **60**, 586–595.

Boucias, D.G. and Pendland, J.C. (1998) *Principles of Insect Pathology*, Kluwer Academic, Boston, MA.

Briggs, C.J. (1993) Competition among parasitoid species on a stage-structured host and its effect on host suppression. *Am. Nat.*, **141**, 372–397.

Brobyn, P.J., Clark, S.J. and Wilding, N. (1988) The effect of fungus infection of *Metopolophium dirhodum* (Hom.: Aphididae) on the oviposition behavior of the aphid parasitoid *Aphidius rhopalosiphi* (Hym.: Aphidiidae). *Entomophaga*, **33**, 333–338.

Brooks, W.M. (1993) Host–parasitoid–pathogen interactions. In N.E. Beckage, S.N. Thompson and B.A. Federici (eds.), *Parasites and Pathogens of Insects*, vol. 2: *Pathogens*, Academic Press, San Diego, CA, pp. 231–272.

Caballero, P., Vargas-Osuna, E. and Santiago-Alverez, C. (1991) Parasitization of granulosis-virus infected and noninfected *Agrotis segetum* larvae and the virus transmission by three hymenopteran parasitoids. *Entomol. Exp. Appl.*, **58**, 55–60.

Carruthers, R.I., Larkin, T.S. and Firstencel, H. (1992) Influence of thermal ecology on the mycosis of a rangeland grasshopper. *Ecology*, **73**, 190–204.

Cloutier, C. and Jean, C. (1998) Synergism between natural enemies and biopesticides: a test case using the stinkbug, *Perillus bioculatus* (Hemiptera: Pentatomidae) and *Bacillus thuringiensis tenebrionis* against Colorado potato beetle (Coleoptera: Chrysomelidae). *J. Econ. Entomol.*, **91**, 1096–1108.

Dahlman, D.L. and Vinson, S.B. (1993) Teratocytes: developmental and biochemical characteristics. In N.E. Beckage, S.N. Thompson and B.A. Federici (eds.), *Parasites and Pathogens of Insects*, vol. 1, *Parasites*, Academic Press, New York, pp. 145–165.

DeBach, P. and Rosen, D. (1991) *Biological Control by Natural Enemies*. Cambridge University Press, Cambridge, UK.

El-Sufty, R. and Führer, E. (1981) Wechselbeziehungen zwischen *Pieris brassicae* L. (Lep., Pieridae), *Apanteles glomeratus* L. (Hym., Braconidae) und dem Pliz *Beauveria bassiana* (Bals.) Vuill. *Z. Angew. Entomol.*, **92**, 321–329.

El-Sufty, R. and Führer, E. (1985) Wechselbeziehungen zwischen *Cydia pomonella* L. (Lep., Tortricidae), *Ascogaster quadridentatus* Wesm. (Hym., Braconidae) und dem Pilz *Beauveria bassiana* (Bals.) Vuill. *Z. Angew. Entomol.*, **99**, 504.

Entwistle, P.F. (1982) Passive carriage of baculoviruses in forests. In *Invertebrate Pathology and Microbial Control, Proc. 3rd Int. Congr. Invertebr. Pathol.*, Brighton, UK, pp. 344–351.

Entwistle, P.F. and Evans, H.F. (1985) Viral control. In G.A. Kerkut and L.I. Gilbert (eds.), *Comprehensive Insect Physiology, Biochemistry and Pharmacology*: vol. 12, *Insect Control*, Pergamon Press, Oxford, pp. 347–412.

Escribano, A., Williams, T., Goulson, D., Cave, R.D. and Caballero, P. (2000) Parasitoid–pathogen–pest interactions of *Chelonus insularis, Campoletis sonorensis*, and a nucleopolyhedrovirus in *Spodoptera frugiperda* larvae. *Biol. Cont.*, **19**, 265–273.

Feener, D.H. and Brown, B.V. (1997) Diptera as parasitoids. *Annu. Rev. Entomol.*, **42**, 73–97.

Finney, D.J. (1971) *Probit Analysis*, Cambridge University Press, Cambridge, UK.

Flexner, J.L., Lighthart, B. and Croft, B.A. (1986) The effects of microbial pesticides on non-target, beneficial arthropods. *Agric. Ecosyst. Environ.*, **16**, 203–254.

Fransen, J.J. and van Lenteren, J.C. (1993) Host selection and survival of the parasitoid *Encarsia formosa* on greenhouse whitefly, *Trialeurodes vaporariorum*, in the presence of hosts infected with the fungus *Aschersonia aleyrodis*. *Entomol. Exp. Appl.*, **69**, 239–249.

Fransen, J.J. and van Lenteren, J.C. (1994) Survival of the parasitoid *Encarsia formosa* after treatment of parasitized greenhouse whitefly larvae with fungal spores of *Aschersonia aleyrodis*. *Entomol. Exp. Appl.*, **71**, 235–243.

Franz, J.M. and Krieg, A. (1957) Virosen europäischer forstinsekten. *Z. Pflanzenkr. Pflanzenschutz*, **64**, 1–9.

Führer, E. and El-Sufty, R. (1979) Production of fungistatic metabolites by teracytes of *Apanteles glomeratus* L. (Hym., Braconidae). *Z. Parasitenkd.*, **59**, 21–25.

Führer, E. and Willers, D. (1986) The anal secretion of the endoparasitic larva *Pimpla turionellae*: sites of production and effects. *J. Insect Physiol.*, **32**, 361–367.

Furlong, M.J. and Pell, J.K. (1996) Interactions between the fungal entomopathogen *Zoophthora radicans* Brefeld (Entomophthorales) and two hymenopteran parasitoids attacking the diamondback moth, *Plutella xylostella*. *J. Invertebr. Pathol.*, **68**, 15–21.

Glare, T.R. (1994) Stage-dependent synergism using *Metarhizium anisopliae* and *Serratia entomophila* against *Costelytra zealandica*. *Biocont. Sci. Technol.*, **4**, 321–329.

Godwin, P.A. and Shields, K.S. (1984) Effects of *Blepharipa pratensis* (Dip.: Tachinidae) on the pathogenicity of nucleopolyhedrosis virus in stage V of *Lymantria diapar* (Lep.: Lymantriidae). *Entomophaga*, **29**, 381–386.

Gösswald, K. (1934) Physiologische Untersuchungen über die einwirkung ökologischer faktoren, besonders temperatur und luftfeuchtigkeit, auf die entwicklung von *Diprion (Lophyrus) pini* L. zur feststellung der ursachen des massenwechsels. *Z. Angew. Entomol.*, **22**, 331–384.

Greco, W., Unkelbach, H-D., Pöch, G., Sühnel, J., Kundi, M. and Bödeker, W. (1992) Consensus on concepts and terminology for combined-action assessment: the Saariselkä Agreement. *Arch. Complex Environ. Studies*, **4**, 65–69.

Gröner, A. (1990) Safety to nontarget invertebrates of baculoviruses. In M. Laird, L.A. Lacey and E.W. Davidson (eds.), *Safety of Microbial Insecticides*, CRC Press, Boca Raton, FL, pp. 135–147.

Hanski, I. (1981) Coexistence of competitors in patchy environment with and without predation. *Oikos*, **37**, 306–312.

Harper, J.D. (1986) Interactions between baculoviruses and other entomopathogens, chemical pesticides, and parasitoids. In R.R. Granados and B.A. Federici (eds.), *The Biology of Baculoviruses*, CRC Press, Boca Raton, FL, pp. 133–155.

Hochberg, M.E. (1989) The potential role of pathogens in biological control. *Nature*, **337**, 262–265.

Hochberg, M.E. (1991a) Intra-host interactions between a braconid endoparasitoid, *Apanteles glomeratus*, and a baculovirus for larvae of *Pieris brassicae*. *J. Anim. Ecol.*, **60**, 51–63.

Hochberg, M.E. (1991b) Extra-host interactions between a braconid endoparasitoid, *Apanteles glomeratus* and a baculovirus for larvae of *Pieris brassicae*. *J. Anim. Ecol.*, **60**, 65–77.

Hochberg, M.E. (1996) Consequences for host population levels of increasing natural enemy species richness in classical biological control. *Am. Nat.*, **147**, 307–318.

Hochberg, M.E., Hassell, M.P. and May, R.M. (1990) The dynamics of host–parasitoid–pathogen interactions. *Am. Nat.*, **135**, 74–94.

Hurlbert, S.H. (1984) Pseudoreplication and the design of ecological field experiments. *Ecol. Monogr.*, **54**, 187–211.

Inglis, G.D., Johnson, D.L. and Goettel, M.S. (1996) Effects of temperature and thermoregulation on mycosis by *Beauveria bassiana* in grasshoppers. *Biol. Cont.*, **7**, 131–139.

Ives, A.R. and May, R.M. (1985) Competition within and between species in a patchy environment: relation between microscopic and macroscopic models. *J. Theor. Biol.*, **115**, 65–92.

Jaques, R.P. and Morris, O.N. (1981) Compatibility of pathogens with other methods of pest control and with different crops. In H.D. Burges (ed.), *Microbial Control of Pests and Plant Diseases 1970–1980*, Academic Press, New York, pp. 695–715.

Jones, W.A. and Poprawski, T.J. (1996) *Bemisia argentifolii* parasitized by *Eretmocerus* sp. is immune to infection by *Beauveria bassiana*. In T.J. Henneberry, N.C. Toscano, R.M. Faust and J.R. Coppedge (eds.), *Silverleaf Whitefly (Formerly Sweetpotato Whitefly, Strain B): 1996 Supplement to the 5-Year National Research and Action Plan*, USDA, ARS-1996-01, p. 118.

Kaya, H.K. (1982) Parasites and predators as vectors of insects diseases. *Proc. Int. Colloq. Invertebr. Pathol.*, 3rd, Brighton, UK, pp. 39–44.

Kelsey, J.M. (1962) Interaction of virus and insect parasites of *Pieris rapae* L. *Proc. Int. Congr. Entomol.*, **11**, 790–796.

King, E.G. and Bell, J.V. (1978) Interactions between a braconid, *Microplitis croceipes*, and a fungus, *Nomuraea rileyi*, in laboratory-reared bollworm larvae. *J. Invertebr. Pathol.*, **31**, 337–340.

Koppenhöfer, A.M. and Kaya, H.K. (1997) Additive and synergistic interaction between entomopathogenic nematodes and *Bacillus thuringiensis* for scarab grub control. *Biol. Cont.*, **8**, 131–137.

Laigo, F.M. and Paschke, J.D. (1968) *Pteromalus puparum* L. parasites reared from granulosis and microsporidiosis infected *Pieris rapae* L. crysalids. *Philipp Agric.*, **52**, 430–439.

Laird, M., Lacey, L.A. and Davidson, E.W. (eds.) (1990) *Safety of Microbial Insecticides*, CRC Press, Boca Raton, FL.

Lavine, M. and Beckage, N. (1995) Polydnaviruses: potent mediators of host insect immune dysfunction. *Parasitol. Today*, **111**, 368–377.

Lopez, R. and Ferro, D.N. (1995) Larviposition response of *Myiopharus doryphorae* (Diptera: Tachinidae) to Colorado potato beetle (Coleoptera: Chrysomelidae) larvae treated with lethal and sublethal doses of *Bacillus thuringiensis* Berliner subsp. *tenebrionis*). *J. Econ. Entomol.*, **88**, 870–874.

Mesquita, A.L.M. and Lacey, L.A. (2001) Interactions among the entomopathogenic fungus, *Paecilomyces fumosoroseus* (Deuteromycotina: Hyphomycetes), the parasitoid, *Aphelinus asychis* (Hymenoptera: Aphelinidae), and their aphid host. *Biol. Cont.*, **22**, 51–59.

Mohamed, M.A., Coppel, H.C., Hall, D.J. and Podgwaite, J.D. (1981) Field release of virus-sprayed adult parasitoids of the European pine sawfly (Hymenoptera: Diprionidae) in Wisconsin. *Great Lakes Entomol.*, **14**, 177–178.

Niklas, O.F. (1939) Zum massenwechsel der tachine *Parasetigena segregata* Rond. (*Phorocera agilis* R.-D.) in der Rominter Heide. (Die parasitierung der nonne durch insekten. Teil II). *Z. Angew. Entomol.*, **26**, 63–103.

Onstad, D.W. and Kornkven, E.A. (1999) Persistence of natural enemies of weeds and insect pests in heterogenous environments. In B.A. Hawkins and H.V. Cornell (eds.), *Theoretical Approaches to Biological Control*, Cambridge University Press, Cambridge, UK, pp. 349–367.

Pell, J.K. and Vandenberg, J.D. (2002) Interactions among the Russian wheat aphid, *Paecilomyces fumosoroseus* and the convergent ladybird. *Biocont. Sci. Technol.* (In press).

Pöch, G., Reiffenstein, R.J., Köck, P. and Pancheva, S.N. (1995) Uniform characterization of potentiation in simple and complex situations when agents bind to different molecular sites. *Can. J. Physiol. Pharmacol.*, **73**, 1574–1581.

Robertson, J.L. and Preisler, H.K. (1992) *Pesticide Bioassay with Arthropods*, CRC Press, Boca Raton, FL.

Rosenheim, J.A., Kaya, H.K., Ehler, L.E., Marois, J.J. and Jaffee, B.A. (1995) Intraguild predation among biological-control agents: theory and evidence. *Biol. Cont.*, **5**, 303–335.

Roy, H.E. and Pell, J.K. (2000) Interactions between entomopathogenic fungi and other natural enemies: implications for biological control. *Biocont. Sci. Technol.*, **10**, 737–752.

Roy, H.E., Pell, J.K., Clark, S.J. and Alderson, P.G. (1998) Implications of predator foraging on aphid pathogen dynamics. *J. Invertebr. Pathol.*, **71**, 236–247.

Roy, H.E., Pell, J.K. and Alderson, P.G. (2001) Targeted dispersal of the aphid pathogenic fungus *Erynia neoaphidis* by the aphid predator *Coccinella septempunctata*. *Biocont. Sci. Technol.*, **11**, 99–110.

Sait, S.M., Begon, M., Thompson, D.J. and Harvey, J.A. (1996) Parasitism of baculovirus-infected *Plodia interpunctella* by *Venturia canescens* and subsequent virus transmission. *Funct. Ecol.*, **10**, 586–591.

Salama, H.S. and Zaki, F.N. (1983) Interaction between *Bacillus thuringiensis* Berliner and the parasites and predators of *Spodoptera littoralis* in Egypt. *Z. Angew. Entomol.*, **95**, 425–429.

Santiago-Alvarez, C. and Caballero, P. (1990) Susceptibility of parasitized *Agrotis segetum* larvae to a granulosis virus. *J. Invertebr. Pathol.*, **56**, 128–131.

Sokal, R.R. and Rohlf, F.J. (1995) *Biometry*, W.H. Freeman, San Francisco, CA.

Stark, D.M., Mills, N.J. and Purcell, A.H. (1999) Interactions between the parasitoid *Ametadoria misella* (Diptera: Tachinidae) and the granulovirus of *Harrisina brillians* (Lepidoptera: Zygaenidae). *Biol. Cont.*, **14**, 146–151.

Tanada, Y. (1964) Epizootiology of insect diseases. In P. DeBach (ed.), *Biological Control of Insect Pests and Weeds*, Reinhold, New York, pp. 548–578.

Tanada, Y. and Kaya, H.K. (1993) *Insect Pathology*, Academic Press, San Diego, CA.

Teakle, R.E., Jensen, J.M. and Mulder, J.C. (1985) Susceptibility of *Heliothis armigera* (Lepidoptera: Noctuidae) on sorghum to nuclear polyhedrosis virus. *J. Econ. Entomol.*, **78**, 1373–1378.

Temerak, S.A. (1980) Detrimental effects of rearing a braconid parasitoid on the pink borer larvae inoculated by different concentrations of the bacterium, *Bacillus thuringiensis* Berliner. *Z. Angew. Entomol.*, **89**, 315–319.

Temerak, S.A. (1982) Wirkungen zwischen *Bacillus thuringiensis* Berl. und larven der Schlupfwespe *Bracon brevicornis* Wesm. in bzw. an den Raupen von *Sesamia cretica* Led. bei verschiedenen Temeraturen. *Anz. Schaedlingskd. Pflanzenschutz, Umweltschutz*, **55**, 137–140.

Thurston, G.S., Kaya, H.K. and Gaugler, R. (1994) Characterizing the enhanced susceptibility of milky disease-infected scarabaeid grubs to entomopathogenic nematodes. *Biol. Cont.*, **4**, 67–73.

van Lenteren, J.C., van Roermund, H.J.W. and Sütterlin, S. (1996) Biological control of greenhouse whitefly (*Trialeurodes vaporariorum*): how does it work? *Biol. Cont.*, **6**, 1–10.

Vandenberg, J.D., Ramos, M. and Altre, J.A. (1998) Dose-response and age- and temperature-related susceptibility of the diamondback moth (Lepidoptera: Plutellidae) to two isolates of *Beauveria bassiana* (Huphomycetes: Moniliaceae). *Environ. Entomol.*, **27**, 1017–1021.

S.P. Wraight

Versoi, P.L. and Yendol, W.G. (1982) Discrimination by the parasite, *Apanteles melanoscelus*, between healthy and virus-infected gypsy moth larvae. *Environ. Entomol.*, **11**, 42–45.

Vey, A. and Fargues, J. (1977) Histological and ultrastructural studies of *Beauveria bassiana* infection in *Leptinotarsa decemlineata* larvae during ecdysis. *J. Invertebr. Pathol.*, **30**, 207–215.

Vinson, S.B. (1990) Potential impact of microbial insecticides on beneficial arthropods in the terrestrial environment. In M. Laird, L.A. Lacey and E.W. Davidson (eds.), *Safety of Microbial Insecticides*, CRC Press, Boca Raton, FL, pp. 43–64.

Wallner, W.E., Dubois, N.R. and Grinberg, P.S. (1983) Alteration of parasitism by *Rogas lymantriae* (Hymenoptera: Braconidae) in *Bacillus thuringiensis*-stressed gypsy moth (Lepidoptera: Lymantriidae) hosts. *J. Econ. Entomol.*, **76**, 275–277.

Weiser, J. and Veber, J. (1957) Die mikrosporidie *Thelohania hyphantriae* Weiser des weissen bärenspinners und anderer mitglieder seiner biocönose. *Z. Angew. Entomol.*, **40**, 55–70.

Weseloh, R.M. and Andreadis, T.G. (1982) Possible mechanism for synergism between *Bacillus thuringiensis* and the gypsy moth (Lepidoptera: Lymantriidae) parasitoid, *Apanteles melanoscelus* (Hymenoptera: Braconidae). *Ann. Entomol. Soc. Am.*, **75**, 435–438.

Weseloh, R.M., Andreadis, T.G., Moore, R.E.B., Anderson, J.F., Dubois, N.R. and Lewis, F.B. (1983) Field confirmation of a mechanism causing synergism between *Bacillus thuringiensis* and the gypsy moth parasitoid, *Apanteles melanoscelus*. *J. Invertebr. Pathol.*, **41**, 99–103.

Willers, D., Lehmann-Danzinger, H. and Führer, E. (1982) Antibacterial and antimycotic effect of a newly discovered secretion from larvae of an endoparasitic insect, *Pimpla turionella* L. (Hym.). *Arch. Microbiol.*, **133**, 225–229.

Wootton, J.T. (1994) Predicting direct and indirect effects: and integrated approach using experiments and path analysis. *Ecology*, **75**, 151–165.

Wraight, S.P. and Carruthers, R.I. (1999) Production, delivery, and use of mycoinsecticides for control of insect pests of field crops. In F.R. Hall and J.J. Menn (eds.), *Methods in Biotechnology*, vol. 5: *Biopesticides: Use and Delivery*, Humana Press, Totowa, NJ, pp. 233–269.

Young, S.Y. and Yearian, W.C. (1990) Transmission of nuclear polyhedrosis virus by the parasitoid *Microplitis croceipes* (Hymenoptera: Braconidae) to *Heliothis virescens* (Lepidoptera: Noctuidae) on soybean. *Environ. Entomol.*, **19**, 251–256.

Biological Control of
Weeds Using Exotic Insects   9

R.E. Cruttwell McFadyen
Queensland Department of Natural Resources and Mines,
Sherwood Qld, Australia

Introduction

Biological control of weeds falls between entomology and weed science. Although use of plant pathogens for weed control is increasing, weed biocontrol practitioners are still overwhelmingly entomologists. They publish in entomological or ecological journals, but present their results at Weed Conferences in sessions dedicated to biological control, rather than Entomology Conferences, where biological control sessions can be dominated by arthropod pest control. As a result, biocontrol of weeds is often not seen as part of integrated pest management (IPM), and has followed a somewhat different track from biocontrol of arthropod pests: host-testing is given greater importance, and classical biocontrol dominates over IPM.

The economic and environmental importance of weeds is not always appreciated. Herbicides make up 47 percent of the world agrochemical sales, with insecticides next at 29 percent (Woodburn, 1995). In the USA, herbicides were 68 percent of total pesticide production in 1993 (Duke, 1997). In the developing world, weeding, usually by hand, accounts for up to 60 percent of total preharvest labor input (Webb and Conroy, 1995). If uncontrolled, weeds can cause complete yield loss, a record equalled by few insect pests or pathogens. Invasive weeds cause enormous environmental damage, which is only now being recognized (see later).

Nordlund (1996) reviewed the different concepts of biocontrol as applied to weeds and insects, and in particular the different "conceptual models" of biocontrol used by entomologists and plant pathologists. Following him, I use DeBach's (1964) definition of biological control as "the actions of parasites, predators, and pathogens in maintaining another organism's density at a lower average than would occur in their absence." Within this definition, there are three different techniques for applied biocontrol: (i) conservation — protection or maintenance of existing populations of biocontrol agents; (ii) augmentation — regular action to increase populations of biocontrol agents, either by periodic releases or by environmental manipulation; and (iii) classical biocontrol — the importation and release of exotic biocontrol agents, with the expectation that the agents will become established and further releases will not be necessary. Unlike the biocontrol of insect pests, where IPM primarily relies on conservation and augmentation, classical biocontrol is the mainstay of weed biological control, and conservation is hardly used (Harris, 1993). Augmentation is occasionally used with mycoherbicides and some insects, and in the deliberate use of grazing animals for weed control (Popay and Field, 1996).

Nonclassical biocontrol of weeds

The use of fungi as bioherbicides is an example of augmentation as a biocontrol technique for weeds. There is an extensive literature on the potential of this technique, but little actual use as commercial or practical methods in the field. Particularly in the USA, weed scientists often use the term "biocontrol" to refer solely to the use of pathogens as mycoherbicides, ignoring classical weed biocontrol (Rayachhetry *et al.*, 1996; Bewick, 1996). Despite this, actual use in the field is minimal, with problems with mass production, formulation, and commercialization continuing to prevent wider use (Auld and Morin, 1995; Morin, 1996). As practical, economically viable alternatives to chemical or mechanical weed control, bioherbicides are still very much a nonstarter.

There are a few examples where native insects are artificially increased or otherwise manipulated for the control of native weeds. Native coccids, *Austrotachardia* sp. and *Tachardia* sp., are used for control of *Cassinia* spp., native woody shrubs, in Australia (Holtkamp and Campbell, 1995). Augmentation and preservation (the cessation of insecticide spraying to control grasshoppers and caterpillars) of a native root- and leaf-feeding weevil, *Cleonidius trivittatus*, has been proposed for the control of the native weed purple locoweed (*Astralagus mollissimus*) in the USA (Pomerinke *et al.*, 1995). Conservation/augmentation of the stem-boring agromyzid *Phytomyza orobanchia* Kaltenbach (through collecting infested stalks in autumn, preserving them through winter, and placing them in the fields in early spring) has been used to control the parasitic weeds *Orobanche* spp. in the southern USSR and has been proposed for Morocco and Egypt (Kroschel and Klein, 1999).

Augmentation of introduced biocontrol agents is more widely used, chiefly when the agent dispersal capacity is poor and the weed occurs in discrete scattered areas. Cacti in Australia and South Africa are controlled through the regular redistribution of mealybugs into isolated infestations (Hosking *et al.*, 1988; Moran and Zimmerman, 1991). In Australia, the floating fern salvinia (*Salvinia molesta*) is controlled in ponds and other water bodies by the salvinia weevil, *Cyrtobagous salviniae* Calder & Sands, supplied in bags of infested salvinia for release into the affected ponds (R. Wood, Brisbane City Council, personal communication). The management of water weeds in the USA relies heavily on the manipulative use of biocontrol agents, and special information packages are used to train operational personnel in the procedures (Grodowitz *et al.*, 1996).

Classical biocontrol

Biocontrol of weeds using exotic insects has a long history, since the first programs against lantana in the early 1900s and against prickly pear cactus in the 1920s (Julien and Griffiths, 1998). Partly as a result of the early successes in control of rangeland weeds, and partly because the use of herbicides was so successful against crop weeds in the developed world, the use of biocontrol of weeds has tended to be concentrated on rangeland and environmental weeds and hence in countries with large areas of rangeland. Consequently, the five most active countries, in numbers of weed species targeted and agents released, are the USA, Australia, South Africa, Canada, and New Zealand, in that order, with the USA and Australia nearly twice as active as the others. All these countries have a long history of successful weed biocontrol. For example, Hawaii started in 1902, and has a success rate close to 50 percent, with 7 out of 21 weed species targeted under "complete" control, and significant partial control of three more (Markin *et al.*, 1992; Gardner *et al.*, 1995). There is an increased emphasis now on using biocontrol for weeds of natural ecosystems (here called environmental weeds), which are having a major impact on native ecosystems in Hawaii (Markin *et al.*, 1992). Hawaii undertakes its own foreign exploration programs, and increasingly introduces pathogens as well as insects (Gardner *et al.*, 1995).

The continental USA is actively involved in several programs (Goeden, 1993). Overseas surveys and testing are undertaken through various USDA-ARS laboratories or through CABI Bioscience. Canada has an active weed biocontrol program (Harris, 1993) and usually employs CABI Bioscience for overseas surveys (Gassmann and Schroeder, 1995). For obvious reasons, Canada and the USA work closely together in both overseas exploration and introductions.

Australia is the second most active country, with a long history of introductions starting in 1914 (Julien and Griffiths, 1998). Some current programs are reviewed by Briese (2000). Foreign exploration is usually undertaken by Australian scientists based overseas (McClay *et al.*, 1995; Harley *et al.*, 1995; Goeden and Palmer, 1995; Scott, 1996) or by employing CABI Bioscience, particularly for pathogens which are increasingly used (Evans and Tomley, 1996). New Zealand has several programs under way (Julien and Griffiths, 1998), cooperating with Australia for many and undertaking its own overseas research or contracting CABI Bioscience for others.

South Africa has a very active program, comprehensively reviewed recently (Hoffmann, 1991, 1995). South Africa carries out its own overseas exploration and cooperates closely with Australia, both for shared weed problems and because many plants from each country have become weeds in the other.

Other countries involved in classical biocontrol are Malaysia (Ooi, 1992), Thailand, India (Hirose, 1992), Indonesia, Vietnam, Papua New Guinea, and China. In Africa, Uganda, Zambia, Tanzania, Kenya, Ghana, Côte d'Ivoire, and Benin have active biocontrol projects (Julien *et al.*, 1996). The program against annual ragweed in Russia, the Ukraine, and Croatia (Reznik, 1996) is the only example of biocontrol of weeds in Europe including Britain (Fowler, 1993). There have been recent initiatives toward biocontrol of some crop weeds in Europe, with studies on five major weeds by working groups representing 11 countries (Scheepens *et al.*, 2001), and proposals for the biocontrol of other introduced weeds such as *Solidago altissima* (Jobin *et al.*, 1996).

Classical biocontrol depends on the introduction of exotic insects and as such is subject to legislative control. Where there is a long history of biocontrol, the legislative system is well developed and generally understood and accepted by scientists, government, and even the public. The current system in Australia is well reviewed by Paton (1995). The Australian Quarantine Inspection Service (AQIS) administers the issuing of permits according to protocols developed over the years, and with the assistance of reviewers in each of the eight States and Territories. The Biological Control Act (1984) was designed to deal with conflicts of interest such as that over the biological control of Patterson's curse, *Echium plantagineum* (Cullen and Delfosse, 1985), and provides a legal basis for introduction of control agents (Harris, 1993).

Harris (1993) has reviewed legislation controlling weed biocontrol in Canada and the USA. As with Australia, the legislation was written in broad terms many years ago, and it is the regulations and guidelines that govern application in practice.

In countries where weed biocontrol is infrequently practiced, the lack of any agreed protocols for introductions, or any defined authority to grant permits, can be a major problem, particularly where classical biocontrol of arthropods is also rare. Usually weed biocontrol can follow the procedure existing for insect biocontrol, as in Indonesia (S. Tjitrosoemito, personal communication). The FAO has developed an International Code for the Import and Release of Exotic Biological Control Agents (Labrada, 1996), for biocontrol of arthropod pests as well as weeds. By clarifying procedures and responsibilities, the Code is particularly helpful for countries without a tradition of biocontrol, and was approved by Member States in 1995.

Procedures

Choice of target weeds

Decisions on which weeds are suitable targets for biocontrol programs are based on the benefits to be achieved plus estimates of the probability of success (Wapshere *et al.*, 1989; Peschken and McClay, 1995; Palmer and Miller, 1996). The more widespread and damaging the weed, the greater the potential benefits, but costs and benefits may be hard to quantify for environmental weeds (Harley *et al.*, 1995;

Blossey *et al.*, 1996; Fowler *et al.*, 1996). Whether a biocontrol program will be successful depends on three main factors (Cullen, 1995):

- The damage each individual agent can do to the plant
- The ecology of the agent, which determines the population density achieved in the new environment
- The ecology of the weed, which determines whether the total damage is significant in reducing its population

The first is relatively easy to determine; the problem is to predict the other two, and most of the numerous predictions made over the last 40 years have been proved wrong (McFadyen, 1998, 2000). About the only valid prediction is that successful biocontrol in one country greatly increases the chances of success in another (Peschken and McClay, 1995). However, there are examples where successful control in one country was not repeated in others (Crawley, 1990; Kelly and McCallum, 1995; Jupp and Cullen, 1996; Woodburn, 1996). Prior use elsewhere also reduces the cost of a biocontrol program, as the expensive overseas survey and testing have already been done (Doeleman, 1989; Cilliers, 1991). Biocontrol is also more difficult where the weed has close relatives of economic or native value, because agents selected must be monophagous, i.e., must feed only on a single species, which reduces the available pool (Olckers *et al.*, 1995).

Serious conflicts of interest, arising where a plant is a weed in one situation and a valuable plant in another, may prevent the use of biocontrol. Where a plant is a serious weed in natural ecosystems but is valuable in other contexts, payment of compensation may be an acceptable solution if the economic value of the plant is minor, for example, strawberry guava, *Psidium cattleianum*, and ginger, *Hedychium gardnerianum*, in Hawaii (Gardner *et al.*, 1995). Where the economic value is great, for example, *Pinus* species, which are major environmental weeds in several countries (Richardson and Higgins, 1997), biocontrol may be inappropriate. Restricting programs to the use of seed predators to reduce spread may be an acceptable alternative, and has been used in South Africa (Dennill and Donnelly, 1991).

Conflict over biocontrol of alien weeds in natural ecosystems may arise from concern with maintaining or reestablishing the indigenous ecosystem. Some conservation groups argue that no biocontrol agent should be released until all potential effects are fully understood, but this would in practice mean no further biocontrol of weeds in natural areas, and it is generally agreed that some risks must be taken. Where the weed invasion has been rapid and recent, the harmful effects are obvious and there is widespread support for biocontrol, for example, of giant sensitive plant, *Mimosa pigra*, in Kakadu National Park in northern Australia or rubbervine, *Cryptostegia grandiflora*, in riverine ecosystems in north Queensland (Humphries *et al.*, 1991). Where the invasion took place gradually over a long period, changes in the ecosystem are gradual and the alien weed may become accepted as part of the ecosystem, which can cause conflict when biocontrol is proposed. Where salt cedar *Tamarix* spp. has replaced native *Salix* spp. in the USA, endangered birds use the salt cedar for nesting. This is seen as a reason to refuse biocontrol permits because of a belief that biocontrol will suddenly and totally eradicate the plant in question, leaving the birds with nowhere to nest or feed (DeLoach *et al.*, 1996). In practice, biocontrol never eradicates the target weed, and timetables of 10 years or more for successful control are usual, during which time native vegetation can gradually replace the alien plant. There is a good discussion of these issues in relation to the introduced weed heather (*Calluna vulgaris*) in New Zealand (Syrett, 1995).

Agent selection

The steps involved in a weed biocontrol program (Wapshere *et al.*, 1989; Harley and Forno, 1992), after the initial decision that classical biocontrol is appropriate, are overseas exploration, selection and testing of agents, rearing and release, and evaluation. Overseas exploration requires correct identification of the weed and its country of origin, which is not always straightforward (Wapshere *et al.*, 1989; Gassmann and Schroeder, 1995). Genetic analysis, based on specific plant chemicals (Holden and Mahlberg, 1995), isozymes (Lanaud *et al.*, 1991), and DNA (Nissen *et al.*, 1995; Radford *et al.*, 2000), is being used to identify and characterize the different strains of a weed. This also facilitates the collection of agents from the correct strain and locality.

Agent selection is the critical step, and choice of the best agent is the "holy grail" of weed biocontrol. In 1991, each agent tested and introduced cost US$400 000, or three scientist-years (Schroeder and Goeden, 1986; Harris, 1991), which, with technical support and facilities, would be about US$500 000 in 2001. Over the years, there have been many theories or protocols on how to choose the best agent (Harris, 1991; Goeden, 1993). Protocols for agent selection, although useful discussion points, have proved of little or no predictive value (Schroeder and Goeden, 1986; Blossey, 1995a; Gassmann and Schroeder, 1995). Sometimes the "best" agent turns out not as good as expected (Jupp and Cullen, 1996; Woodburn, 1996), or an insect may perform better than expected (Hoffmann and Moran, 1995). The major problem with prediction is that success does not depend on features of the insect as much as upon environmental factors such as climate and the presence of parasites or predators (Gassman and Schroeder, 1995; Blossey, 1995b). Predictions based on prerelease studies of agent impact in their native range (Blossey *et al.*, 1996) may prove equally useless, chiefly because it is impossible to predict the factors affecting agent populations in the new country (McFadyen, 1986; Jupp and Cullen, 1996). Predictions based on climatic analysis need to be treated with caution (McFadyen, 1991), as the best climatic match is no guarantee of success while, conversely, some agents have thrived outside their "normal" climatic range (McFadyen, 1985; Gassmann and Schroeder, 1995). They can, however, be useful in extreme climates such as in Canada (McClay, 1996b).

Host specificity testing

The necessity for detailed host specificity testing of all agents before field release has been an accepted doctrine since the biocontrol of prickly pear (Dodd, 1940), in contrast to arthropod biocontrol (Howarth, 1991). As a result, there have been no "disasters" in the history of weed biocontrol (McFadyen, 1998). Unfortunately, this has not prevented critics from quoting examples taken almost exclusively from biocontrol of arthropods, but including weed biocontrol in the negative statements made (Howarth, 1991; Simberloff and Stiling, 1996), or referring specifically to "notable disasters where organisms were introduced to control weeds with little regard to non-target organisms" without giving examples or references (Cronk and Fuller, 1995). The only nonspecific agents used in weed biocontrol have been fish, introduced into several countries for fishing and to control submerged aquatic weeds (Julien and Griffiths, 1998), with frequently disastrous results.

Over the years, host specificity testing has developed from the testing of long lists of crop plants unrelated to the host weed, to use of targeted lists of plants closely related to the weed and including native plants (Fornasari and Turner, 1995). The aim is no longer to demonstrate that a group of valuable plants will not be attacked, but has become the determination of the potential host range of the agent, and therefore of which plants if any are at risk in the field (Wapshere, 1989; Cullen 1990; Blossey, 1995a; McClay, 1996a). It is usual to test feeding in all mobile stages (adults and mobile larvae or nymphs). Where larvae are immobile, feeding inside the plant or on the roots, adult oviposition choice is tested instead. Because oviposition in itself does not usually cause significant damage, the critical factor is the ability to feed and develop on the test plants (Wapshere, 1989; Cullen, 1990; Blossey, 1995a; McClay, 1996a). Test results are published in entomological and biocontrol journals and in the proceedings of the International Symposia on Biological Control of Weeds. Results for rejected agents are often not published, with some exceptions (McFadyen and Marohasy, 1990b; Palmer and Goeden, 1991), which is regrettable as it can give the impression that potential agents are never rejected.

The phylogeny of both the plant and the insect is critical to host specificity in most groups (Briese, 1996). Understanding of host specificity is greatly improved when the insects attacking a complete taxonomic group of plants are known, for example, for the thistles in Europe (Sheppard *et al.*, 1995; Briese, 1996) or for the Ambrosiinae in North America (Goeden and Palmer, 1995; McClay *et al.*, 1995), or where the host relationships of a taxonomic group of insects are studied (Anderson, 1993; Futuyma *et al.*, 1995). Developing theories on evolution of host specificity are having an impact on host testing and the understanding of host range. It is now generally agreed that evolution in phytophagous insects has been from generalists to specialists, with the result that, in specialists, there may be little surviving genetic variation in ability to utilize different host plants for oviposition or for larval or adult feeding (Futuyma *et al.*, 1995). In other words, highly host-specific insects introduced into a new country are most unlikely to become selected for ability to use novel plants as hosts (Marohasy, 1996).

The major problem encountered by biocontrol researchers is the interpretation of results where feeding occurs in the tests but not in the field. This may often be the result of artificial confinement (Harris and McEvoy, 1995; Balciunas *et al.*, 1996), but it may mean the field data are inadequate. For example, the chrysomelid *Ophraella communa* was believed to be restricted to plants in the subtribe Ambrosiinae of the Heliantheae and was not known from sunflower *Helianthus annuus* or other *Helianthus* spp. In tests, development occurred on sunflower and the species was rejected as a biocontrol agent (Palmer and Goeden, 1991). Subsequently, the results of the laboratory tests were confirmed when *O. communa* was found in the field on *Helianthus ciliaris* and *Ratibida pinnata*, both in the subtribe Helianthiinae. If extensive development or feeding occurs in laboratory tests on plants not attacked in the country of origin, very careful analysis is needed to determine whether other factors might prevent attack on these plants under field conditions. These can be specialized pupation requirements (Balciunas *et al.*, 1996), or aggregation responses to chemicals from the damaged plant or to feces containing these chemicals (Wan and Harris, 1996). If no such limiting factors exist, then it must be assumed that attack will take place. Testing must take into account the possibility that very high population levels developing on the host weed may result in starving insects dispersing onto adjacent crops or other plants, where significant damage may occur even if development or long-term survival is not possible. For this reason, some kind of non-choice test on closely related "at risk" plants must be part of the testing schedule (Wapshere, 1989).

Host testing can never give absolute answers (i.e., guarantee the agent will never attack other plants) but provides the basic information required for a process of risk assessment (McClay, 1996a). When test results indicate that attack will occur on desirable native or crop plants, the decision whether or not to release the agent is ultimately political, where the risks of release are weighed against the consequences of alternative control methods. Agents have been released in the knowledge that they would attack nontarget plants, where the relative value of the nontarget plant was significantly lower than the damage (economic or environmental) being caused by the weed (McFadyen and Marohasy, 1990a; Olckers *et al.*, 1995). In such cases, it is important that resources be allocated for careful evaluation of the actual field impact of the agent on both weed and nontarget plants.

Evaluation

In the past, little follow-up evaluation has been done (Blossey, 1995a), chiefly because financial sponsors took the view that it would be obvious whether or not the weed was successfully controlled. The first step is prerelease studies of the weed in the target country, essential if there is to be adequate evaluation of the impact of biocontrol, but too often these are not done. Even basic weed population ecology may not be known (McFadyen, 1986). Examples of good prerelease studies are those on *Mimosa pigra* in Australia (Lonsdale *et al.*, 1989), on *Cynoglossum* in Canada (DeClerke-Floate, 1996), and on *Chrysanthemoides* in Australia and South Africa (Scott, 1996).

Most evaluation is undertaken after the initial release and establishment. More effort is now going into the distribution of already established agents, and this requires evaluation to justify continued spending on distribution (Blossey *et al.*, 1996). The evaluation may, however, be limited to monitoring the presence and spread of the insect population, without evaluation of impact. In a review of evaluation studies, McClay (1995) emphasized the need to evaluate the weed, not the agent. It is not sufficient to count the agent numbers: evaluation requires a study of the impact on the weed and its population dynamics. Laboratory experiments are commonly undertaken (Spollen and Piper, 1995); these demonstrate that damage affects the plant, but the relationship to field conditions is often hard to demonstrate. Experimental studies in the field, where the biocontrol agent is excluded artificially by insecticide or exclusion cages, are not common, mainly because of the long time required for effects to develop and the resulting need for long-term treatment with insecticides or maintenance of cages (McFadyen, 1998; Dhileepan, 2001). As a minimum, field sites must be established to collect long-term data to measure the weed abundance (plant numbers and seed production or seed bank size) prior to and after release of the biocontrol agents.

Postestablishment analyses have been used to determine reasons for success or failure. For example, temperature and rainfall are the factors behind the successful control of Noogoora burr, *Xanthium occidentale*, by the rust *Puccinia xanthii* in Queensland, and its failure in the Northern Territory and Western Australia (Morin *et al.*, 1996). An evaluation of the unsuccessful control of the thistle *Carduus nutans* in New Zealand 20 years after releases demonstrated why a high level of seed destruction by the biocontrol agent has a minimal impact on the population dynamics of the weed, in contrast to the experience in Canada and the USA (Kelly and McCallum, 1995). However, there is still no clear understanding of the reasons for these differences, and the only solution is to introduce new agents attacking other life stages of the plant.

Results achieved

Successes and failures

Consideration of successes is bedevilled by the problem of assessment — when is control "successful"? The three best definitions are those used by Hoffmann (1995):

- Complete, when no other control method is required or used, at least in areas where the agent(s) are established
- Substantial, where other methods are needed but less effort is required (e.g., less herbicide or less frequent application)
- Negligible, where despite damage inflicted by agents, control of the weed is still dependent on other control measures

"Complete" control does not mean that the weed is eradicated or is no longer a component of the weed flora, but that control measures are no longer required solely against the target weed, and that crop or pasture yield losses can no longer be attributed solely or largely to this weed.

Success rates are generally quoted as 60 percent of agents introduced resulting in successful establishment, and 33 percent of these resulting in control (Crawley, 1989, 1990; Williamson and Fitter, 1996). These analyses are often distorted by the inclusion of data from recent programs where equilibrium has not yet been reached or control achieved. In fact, as agent establishment may take many years (R.E.C. McFadyen unpublished data; Vitelli *et al.*, 1996) and control up to 10 years after that (Hoffmann, 1995), analyses of success rates should discard all programs with less than 10 years elapsed since the last introduction, and treat data from the following 10 years with caution.

Of greater importance is the proportion of programs, rather than individual agents, that achieve successful control. This is 83 percent for South Africa, with 6 weeds out of 23 targeted under complete control and a further 13 under substantial control (Hoffmann, 1995). In Hawaii nearly 50 percent control has been achieved, with 7 weeds out of 21 under complete control, and significant control in three more (Markin *et al.*, 1992; Gardner *et al.*, 1995). In Australia, of 15 completed programs, 12 resulted in complete control, i.e., 80 percent. Of 21 on-going programs commenced before 1986, four have achieved complete control and three substantial and still improving control, i.e., 33 percent, giving an overall success rate of 53 percent. Negative attitudes toward biocontrol may also result from an expectation that the weed must be completely controlled, with partial successes counted as failures, so that the very real savings achieved are not measured or recognized (Hoffmann, 1995; McFadyen, 2000).

In early programs, very small releases of control agents were often made, and establishment rates were consequently poor. Increased effort is now being made to release large numbers of agents, often using field cages initially, and more effort is being put into redistribution of agents once they are established (Blossey *et al.*, 1996). As a result, establishment rates are now approaching 100 percent for some programs (Harley *et al.*, 1995), and improved distribution may also reduce the lag-time between establishment and successful control. There is always a conflict between making several releases to "spread the risk," and making larger releases to increase the viability of the initial populations; ecological theory can help determine the optimum release size for establishment with different agents (Shea and Possingham, 2000). Analysis of past releases also improves understanding of the factors involved in successful establishment (Hight *et al.*, 1995). Similarly, releases of agents that have been laboratory-reared for many generations is risky unless careful precautions have been taken to prevent in-breeding or selection for undesirable characteristics (McFadyen, 2001).

Forty-one weeds have been successfully controlled somewhere in the world using introduced insects and pathogens (McFadyen, 2000). With many of these weeds, the successful control has been repeated in several countries and regions of the world, and the savings to agriculture and the environment are enormous. Economic evaluations of classical biocontrol, for both arthropod and weed pests, have been recently reviewed for Australia (Cullen and Whitten, 1995). The successful biocontrol of noogoora burr in Queensland resulted in annual benefits (in 1991) of US$720 000, a return of 2.3:1 (Chippendale, 1995). Evaluations of the successful control of skeleton weed (Marsden *et al.*, 1980) and of tansy ragwort *Senecio jacobaeae* (Coombs *et al.*, 1995), have demonstrated benefit/cost ratios of 112 and 15. Biocontrol programs also result in substantial noneconomic benefits, in sustainability of the success, and in equity, in that benefits are not limited to those who can afford the product (Doeleman, 1989). Both benefits and costs are particularly hard to determine for environmental weeds, where agricultural costs are not involved (Cullen and Whitten, 1995; Fowler *et al.*, 1996). The program against *Passiflora mollissima* in Hawaii cost about US$1 million over 5 years, though "additional effort beyond 5 years will likely be required for successful control" (Markin *et al.*, 1992). These costs are small in relation to costs of controlling agricultural weeds with herbicides, but may be large in relation to budgets for management of natural ecosystems.

Major successes in the last two decades include tansy ragwort in the USA (McEvoy *et al.*, 1991), and nodding thistle, *Carduus nutans*, in Canada and the USA (Frick, 1978; Harris, 1984). Less well-known examples are the successful control of black sage, *Cordia curassavica*, in Malaysia after the earlier success in Mauritius (Ooi, 1992). Control of the annual weed noogoora burr in Australia (Chippendale, 1995; Morin *et al.*, 1996), of harrisia cactus in Australia, *Harrisia martinii* (McFadyen, 1986), and of the annual weed giant sensitive plant, *Mimosa invisa*, in Australia, Papua New Guinea (Kuniata, 1994; Ablin, 1995), and the Cook Islands are other examples. The perennial shrub *Chromolaena odorata* has been successfully controlled in the Marianas (Siebert, 1989), in Ghana (Timbilla and Braimah, 2002), and now in large areas of northern Sumatra (Desmier de Chenon *et al.*, 2002). The perennial shrubs hamakua pamakani, *Ageratina riparia*, and klamath weed, *Hypericum perforatum*, are now under complete control in Hawaii (Markin *et al.*, 1992; Gardner *et al.*, 1995). The trees *Acacia saligna* and *Sesbania punicea* have been successfully controlled in South Africa (Morris, 1997; Hoffmann and Moran, 1998).

The programs against floating water weeds have been the outstanding success of the last two decades. Three water weeds — water lettuce, *Pistia stratiotes*; water hyacinth, *Eichhornia crassipes*, and salvinia, *Salvinia molesta* — have been successfully controlled in the tropics and subtropics where they have been major weeds (Thomas and Room, 1985; Harley *et al.*, 1990; Julien *et al.*, 1996). Alligator weed, *Alternanthera phileroxoides*, has been successfully controlled in aquatic habitats in subtropical climates (Julien and Chan, 1992). There was early success with water hyacinth in the USA, but implementation of successful biocontrol in other countries has been slow and is only now gaining momentum (Julien and Griffiths, 1998). The program against salvinia was a failure initially, as the agents tested and introduced from northern South America, including the weevil *Cyrtobagous singularis* Hustache, were not successful. Ten years later, new surveys in southern Brazil found the weevil *C. salviniae*, initially thought to be a host-race of the first species but subsequently shown to be a new species with a different biology, which was significantly more damaging to the host plant (Sands and Schotz, 1985). This weevil has since successfully controlled salvinia in all tropical and subtropical countries where it has been introduced (Julien and Griffiths, 1998).

Financial and social benefits from control of water hyacinth and salvinia in particular have been enormous. Because waterways are used for transport as well as fisheries, irrigation, and water supply, an entire society can be disrupted or even destroyed if dense mats of floating water weeds prevent movement between villages. This was the situation caused by salvinia in the Sepik river in Papua New Guinea, and by water hyacinth in the rivers of southern Irian Jaya on the same island, as well as in many other areas of the world. Control by chemical or mechanical means was impossible, as the areas involved were so large and the weed multiplied so fast. It is hard, therefore, to place a monetary value on the successful control of these weeds, but this value must include the continued prosperity and productivity of the Sepik river villages. When the weevil was used to control salvinia in Sri Lanka, the benefit/cost ratio was calculated to be 1675:1 — costs were low because the weevil had already been tested and used in Australia (Doeleman, 1989).

Damage to nontarget plants

There are very few cases of damage to nontarget plants resulting from the introduction of insects for the biocontrol of weeds (McFadyen, 1998). In most of these, at the time when the decision was made for their release, the host range of the insect was known to include genera having one or more plants native to the country of introduction. However, at the time when the releases were made, attack on native plants of no economic value was not seen as a problem. Examples are the attack by the weevil *Rhinocyllus conicus* (Froelich) on native thistles in the USA (Turner *et al.*, 1987), and the attack by the cinnabar moth *Tyria jacobaeae* (Linnaeus) on native *Senecio* species in Canada and the USA (Diehl and McEvoy, 1990). The problem of *Cactoblastis cactorum* (Berg) attacking native cacti in Florida and the southern USA is a somewhat different example in this category (Bennett and Habeck, 1995). When it was introduced into the West Indies in the late 1950s to control the native *Opuntia triacantha*, no value was placed on native cacti and neither conservationists nor the US government raised objections when the release was published (Simmonds and Bennett, 1966). The moth has since spread throughout the Caribbean, both naturally and through deliberate introductions, until in 1989 it was found in the Florida Keys (Pemberton, 1995). Here it is threatening native *Opuntia* spp. endangered by clearing and development of the Keys (Johnson and Stiling, 1996). It is likely to continue its spread westward into Mexico and the cactus country of the southwest USA (Simberloff and Stiling, 1996), where its impact may be severe unless reduced by the effects of parasitism.

The other category of nontarget damage which has been recorded is spillover effects, where large populations develop locally or seasonally on the target weed. When the weed is destroyed or dries off seasonally, the starving insects move onto and damage crops. The beetle *Zygogramma bicolorata* Pallister was released in India to control parthenium weed, but where large beetle populations developed there was substantial though short-term damage to adjacent sunflower crops. This attack on sunflower was not anticipated, despite test results showing that starving beetles will feed on sunflower (McFadyen and McClay, 1981), because of failure to appreciate the impact of the very large populations of the beetle in the initial years after release. As the weed becomes less abundant, the problem is subsiding (Jayanth and Ganga Visalakshy, 1996). In all known examples, economic losses have been very minor and/or temporary, and are far outweighed by the benefits obtained. Environmental damage is harder to assess, and in most cases has not been properly evaluated. Damage due to the agents must be weighed against the benefits from control of these previously widespread introduced weeds, and from the cessation of chemical use over extensive areas of grassland and natural vegetation.

Nevertheless, despite the long history of successful and safe biocontrol of weeds, practitioners do need to recognize the risks involved. Classical biocontrol is irreversible: an agent once widely established in a new country cannot be eradicated, and, therefore, it is essential that all potential consequences be adequately considered beforehand. Furthermore, it needs to be recognized that successful biocontrol agents may disperse far beyond the original target area, for example, lantana seed fly in southeast Asia (Ooi, 1987) and cactoblastis in Florida (Bennett and Habeck, 1995;

R.E. Cruttwell McFadyen

Pemberton, 1995), and their impact in the full potential range must be considered. On the other hand, those who oppose releases until all possible consequences are understood, in effect indefinitely, need to remember that the uncontrolled growth of many weeds is already causing environmental damage and species extinctions, and these will continue so long as control is delayed. Classical biocontrol involves a process of decision analysis — balancing the risks of releasing versus the risks of no control (McClay, 1996a). Accurate prediction of the potential host range of the agent minimizes the risk. Practitioners need to consider the full range of issues involved: the importance of nontarget plants and associated fauna; the probable impact of any damage; as well as the probable damage to nontarget plants from other control options including no control. Appropriate public bodies must make decisions after an open process: the role of biocontrol scientists is to submit proposals and supply information but not to make the final decision. Resources have to be provided for long-term postrelease monitoring of impacts on both target and nontarget species (Howarth, 1991). This will also improve our understanding of host specificity in the field (McClay, 1996a).

Future of weed biocontrol

Plant introductions for forestry and pasture have increased greatly in the last three decades, which will inevitably lead to increased weed problems after a lag-time typically of 50 years (Hughes, 1995). The nursery trade is another problem; most pernicious weed species were deliberately introduced as garden ornamentals (Crawley *et al.*, 1996). In the USA, 85 percent of woody plants invading natural areas were introduced as ornamentals and 14 percent for agricultural use (Reichard, 1997), while 13 percent of "pasture" plants introduced into northern Australia have become weeds (Lonsdale, 1994). Unlike biocontrol agents, these plant introductions are not subject to any controls in most countries (Panetta, 1993). Because of the characteristics selected for — easy establishment, rapid growth, high competitiveness — such introductions are more likely to become invasive weeds.

To compound the problem, damage caused by alien plants to natural ecosystems has been underresearched and its full seriousness is not appreciated (McFadyen, 2002). Alien plant invasions can alter the primary production level and the whole vegetation structure (Cronk, 1995). The impact of invasive weeds on native ecosystems has recently been reviewed for the world as a whole (Cronk and Fuller, 1995) and for the USA (Randall, 1996). Alien plant invasions now affect conservation areas on every continent except Antarctica (Cronk, 1995). As a result, conservation scientists and managers are increasingly accepting that "biocontrol is the only resort when the invasion is 'out of control'" (Cronk and Fuller, 1995; Cronk, 1995), but this understanding has not reached the general conservation community, let alone the public as a whole. Biocontrol of weeds of environmental areas faces difficulties in assessing damage caused, when economic losses are no longer the criteria, and in working with different interest groups and financial sponsors (Fowler and Waage, 1996). However, despite these problems, well-funded programs against nonagricultural weeds are being developed (Harley *et al.*, 1995; Blossey *et al.*, 1996).

Biocontrol will be the only feasible solution to many of these weed problems, as extensive use of herbicides or mechanical control methods in conservation areas would be very damaging as well as prohibitively expensive. The FAO now regards biocontrol of weeds as the major option to be promoted. It is currently supporting programs for the biocontrol of water hyacinth in Latin America and Africa, itchgrass (*Rottboelia* spp.) in Central America and the Caribbean, *C. odorata* in West Africa, and the parasitic weeds *Orobanche* and *Cuscuta* species in North Africa (Labrada, 1996). However, even when biocontrol is widely seen as successful, time may still be wasted in "reinventing the wheel". With water hyacinth in Africa, agencies have required expensive and time-consuming studies before introducing the three proven biocontrol agents. Biocontrol of water hyacinth takes 3–10 years after release and agencies may insist on a "quick-fix" chemical approach, wasting resources and delaying successful control through disruption of the agent populations (Julien *et al.*, 1996).

Conclusions

Biological control of weeds using imported insects and pathogens is safe, environmentally sound, and cost-effective. The successes listed earlier have already saved many millions of dollars in control costs and increased production. Control of the water weeds salvinia and water hyacinth has preserved the lifestyle of entire communities as well as restoring biodiversity to destroyed aquatic ecosystems. To set against this, after nearly 100 years of use, there are very few examples of damage to nontarget plants, none of which has caused serious economic or environmental damage. Invasive weeds are as environmentally damaging as land clearing, but their attack is more insidious, as the loss of native species, both flora and fauna, is not obvious unless someone is out there measuring.

From the start, the introduction of weed biocontrol agents has been strictly controlled and has required evidence that the agents will not damage nontarget organisms. Unfortunately this is not true of the importation of plant species around the world, as ornamentals, crop or pasture plants, or for forestry. Of the plant and animal species deliberately introduced into the USA for agriculture, for sport, or as pets, respectively 2, 50, and 50 percent have become pests (Pimentel, 1995). It is imperative that plant introductions be subject to the same level of control as biocontrol agents, but meanwhile we can expect a massive increase in invasive weed problems as the plants introduced over the last 50–100 years become naturalized and begin to spread. Up to 13 percent of these can be expected to cause serious economic or environmental damage (Lonsdale, 1994) and classical biocontrol using introduced insects is the preferred method to control them. Mechanical or cultural control is not feasible in environmental areas, and widespread use of herbicides is unacceptable on environmental

R.E. Cruttwell McFadyen

and human health grounds, as well as being economically unsustainable. Classical bio-control is the only safe, practical, and economically feasible method that is sustainable in the long term, and it is essential that the importation of insects and pathogens is not made impossible by unrealistic restrictions and ever-increasing demands for prere-lease studies. It would be irresponsible to claim that no risks are involved, but these risks are small and must be weighed against the risks of alternative control methods where ecosystems and livelihoods are being destroyed and the option of doing nothing is actively harmful rather than being "benign neglect."

References

Ablin, M.P. (1995) *Heteropsylla* sp. (Psyllidae) successfully controls pasture infestations of *Mimosa invisa* within three years of release in Australia. In E.S. Delfosse and R.R. Scott (eds.), *Proc. 8th Int. Symp. Biol. Control Weeds*, CSIRO, Melbourne, pp. 435–436.

Anderson, R.S. (1993) Weevils and plants: phylogenetic versus ecological mediation of evolution of host plant associations in Curculioninae (Coleoptera: Curculionidae). *Mem. Entomol. Soc. Can.*, **165**, 197–232.

Auld, B.A. and Morin, L. (1995) Constraints in the development of bioherbicides. *Weed Technol.*, **9**, 638–652.

Balciunas, J.K., Burrows, D.W. and Purcell, M.F. (1996) Comparison of the physiological and realized host-ranges of a biological control agent from Australia for the control of the aquatic weed *Hydrilla verticillata. Biol. Cont.*, **7**, 148–158.

Bennett, F.D. and Habeck, D.H. (1995) *Cactoblastis cactorum*: a successful weed control agent in the Caribbean, now a pest in Florida. In E.S. Delfosse and R.R. Scott (eds.), *Proc. 8th Int. Symp. Biol. Control Weeds*, CSIRO, Melbourne, pp. 21–26.

Bewick, T.A. (1996) Technological advancements in biological weed control with microorganisms: an introduction. *Weed Technol.*, **10**, 600.

Blossey, B. (1995a) A comparison of various approaches for evaluating potential biological control agents using insects on *Lythrum salicaria. Biol. Cont.*, **5**, 113–122.

Blossey, B. (1995b) Host specificity screening of insect biological weed control agents as part of an environmental risk assessment. In H.M.T. Hokkanen and J.M. Lynch (eds.), *Biological Control: Benefits and Risks*, Cambridge University Press, Cambridge, UK, pp. 84–89.

Blossey, B., Malecki, R.A., Schroeder, D. and Skinner, L. (1996) A biological control program using insects against purple loosestrife, *Lythrum salicaria*, in North America. In V.C. Moran and J.H. Hoffmann (eds.), *Proc. 9th Int. Symp. Biol. Control Weeds*, University of Capetown, Stellenbosch, pp. 351–355.

Briese, D.T. (1996) Phylogeny: can it help us to understand host choice by biological weed control agents? In V.C. Moran and J.H. Hoffmann (eds.), *Proc. 9th Int. Symp. Biol. Control Weeds*, University of Capetown, Stellenbosch, pp. 63–70.

Briese, D.T. (2000) Classical biological control. In B. Sindel (ed.), *Australian Weed Management Systems*, R.G. and F.J. Richardson Publications, Melbourne, Australia, pp. 161–192.

Chippendale, J.F. (1995) The biological control of Noogoora burr (*Xanthium occidentale*) in Queensland: an economic perspective. In E.S. Delfosse and R.R. Scott (eds.), *Proc. 8th Int. Symp. Biol. Control Weeds*, CSIRO, Melbourne, pp. 185–192.

Cilliers, C.J. (1991) Biological control of water lettuce, *Pistia stratiotes* (Araceae), in South Africa. *Agric. Ecosyst. Environ.*, **37**, 225–229.

Coombs, E.M., Isaacson, D.L. and Hawkes, R.B. (1995) The status of biological control of weeds in Oregon. In E.S. Delfosse and R.R. Scott (eds.), *Proc. 8th Int. Symp. Biol. Control Weeds*, CSIRO, Melbourne, pp. 463–471.

Crawley, M.J. (1989) Insect herbivores and plant population dynamics. *Annu. Rev. Entomol.*, **34**, 531–564.

Crawley, M.J. (1990) Plant life-history and the success of weed biological control projects. In E.S. Delfosse (ed.), *Proc. 7th Int. Symp. Biol. Control Weeds*, Min. Agric. Forest, Rome, pp. 17–26.

Crawley, M.J., Harvey, P.H. and Purvis, A. (1996) Comparative ecology of the native and alien floras of the British Isles. *Phil. Trans. R. Soc. Lond. B*, **351**, 1251–1259.

Cronk, Q.C.B. (1995) Changing worlds and changing weeds. *Proc. BCPC Symp. Weeds in a Changing World*, **64**, 3–13.

Cronk, Q.C.B. and Fuller, J.L. (1995) *Plant Invaders: The Threat to Natural Ecosystems*, People and Plants Conservation Manuals, WWF International, Chapman and Hall, London, 241 pp.

Cullen, J.M. (1990) Current problems in host-specificity screening. In E.S. Delfosse (ed.), *Proc. 7th Int. Symp. Biol. Control Weeds*, Min. Agric. Forest, Rome, pp. 27–36.

Cullen, J.M. (1995) Predicting effectiveness: fact and fantasy. In E.S. Delfosse and R.R. Scott (eds.), *Proc. 8th Int. Symp. Biol. Control Weeds*, CSIRO, Melbourne, pp. 103–109.

Cullen, J.M. and Delfosse, E.S. (1985) *Echium plantagineum*: catalyst for conflict and change in Australia. In E.S. Delfosse (ed.), *Proc. 6th Int. Symp. Biol. Control Weeds*, Agriculture Canada, Ottawa, pp. 249–292.

Cullen, J.M. and Whitten, M.J. (1995) Economics of classical biological control: a research perspective. In H.M.T. Hokkanen and J.M. Lynch (eds.), *Biological Control: Benefits and Risks*, Cambridge University Press, Cambridge, UK, pp. 270–276.

DeBach, P. (1964) Successes, trends, and future possibilities. In P. DeBach (ed.), *Biological Control of Insect Pests and Weeds*, Chapman and Hall, London, pp. 673–713.

DeClerke-Floate, R. (1996) The role of pre-release studies in developing a biocontrol strategy for hounds's tongue in Canada. In V.C. Moran and J.H. Hoffmann (eds.), *Proc. 9th Int. Symp. Biol. Control Weeds*, University of Capetown, Stellenbosch, pp. 143–148.

DeLoach, C.J., Gerling, D., Fornasari, L., Sobhian, R., Myartseva, S., *et al.* (1996) Biological control programme against saltcedar (*Tamarix* spp.) in the United States of America: progress and problems. In V.C. Moran and J.H. Hoffmann (eds.), *Proc. 9th Int. Symp. Biol. Control Weeds*, University of Capetown, Stellenbosch, pp. 253–260.

Dennill, G.B. and Donnelly, D. (1991) Biological control of *Acacia longifolia* and related weed species (Fabaceae) in South Africa. *Agric. Ecosyst. Environ.*, **37**, 115–135.

Desmier de Chenon, R., Sipayung, A. and Sudharto, P. (2002) A decade of biological control against *Chromolaena odorata* at the Indonesian Oil Palm Research Institute in Marihat. In C. Zachariades, R. Muniappan and L.W. Strathie (eds.), *Proc. Int. Workshop on Biological Control and Management of Chromolaena odorata*, Durban, (in press).

Dhileepan, K. (2001) Effectiveness of introduced biocontrol insects on the weed *Parthenium hysterophorus* (Asteraceae) in Australia. *Bull. Entomol. Res.*, **91**, 167–176.

Diehl, J.W. and McEvoy, P.B. (1990) Impact of the cinnabar moth, *Tyria jacobaeae*, on *Senecio triangularis*, a non-target native plant in Oregon. In E.S. Delfosse (ed.), *Proc. 7th Int. Symp. Biol. Control Weeds*, Min. Agric. Forest, Rome, pp. 119–126.

Dodd, A.P. (1940) *The Biological Campaign against Prickly-Pear*, Commonwealth Prickly Pear Board, Brisbane, 177 pp.

Doeleman, J.A. (1989) Biological control of *Salvinia molesta* in Sri Lanka: an assessment of costs and benefits. *ACIAR Technical Reports 12*, Canberra.

Duke, S.O. (1997) Weed science directions in the USA: what has been achieved and where the USA is going. *Plant Prot. Quart.*, **12**, 2–6.

Evans, H.C. and Tomley, A.J. (1996) Greenhouse and field evaluations of the rubber vine rust, *Maravalia cryptostegiae*, on Madagascan and Australian Asclepiadaceae. In V.C. Moran and J.H. Hoffmann (eds.), *Proc. 9th Int. Symp. Biol. Control Weeds*, University of Capetown, Stellenbosch, pp. 165–169.

Fornasari, L. and Turner, C.E. (1995) Host-specificity of the palearctic weevil *Larinus curtus* (Coleoptera: Curculionidae), a natural enemy of *Centaurea solstitialis* (Asteraceae: Cardueae). In E.S. Delfosse and R.R. Scott (eds.), *Proc. 8th Int. Symp. Biol. Control Weeds*, CSIRO, Melbourne, pp. 385–391.

Fowler, S.V. (1993) The potential for control of bracken in the UK using introduced herbivorous insects. *Pestic. Sci.*, **37**, 393–397.

Fowler, S.V. and Waage, J.K. (1996) New developments, strategies and overviews: synthesis of session 4. In V.C. Moran and J.H. Hoffmann (eds.), *Proc. 9th Int. Symp. Biol. Control Weeds*, University of Capetown, Stellenbosch, pp. 343–347.

Fowler, S.V., Harman, H.M., Memmott, J., Paynter, Q., Shaw, R., *et al.* (1996) Comparing the population dynamics of broom, *Cytisus scoparius*, as a native plant in the United Kingdom and France and as an invasive alien weed in Australia and New Zealand. In V.C. Moran and J.H. Hoffmann (eds.), *Proc. 9th Int. Symp. Biol. Control Weeds*, University of Capetown, Stellenbosch, pp. 19–26.

Frick, K.E. (ed.) (1978) *Biological Control of Thistles in the genus Carduus in the United States*, USDA, 50 pp.

Futuyma, D.J., Keese, M.C. and Funk, D.J. (1995) Genetic constraints on macroevolution: the evolution of host affiliation in the leaf beetle genus *Ophraella*. *Evolution*, **49**, 797–809.

Gardner, D.E., Smith, C.W. and Markin, G.P. (1995) Biological control of alien plants in natural areas of Hawaii. In E.S. Delfosse and R.R. Scott (eds.), *Proc. 8th Int. Symp. Biol. Control Weeds*, CSIRO, Melbourne, pp. 35–40.

Gassmann, A. and Schroeder, D. (1995) The search for effective biological control agents in Europe: history and lessons from leafy spurge (*Euphorbia esula* L.) and cypress spurge (*Euphorbia cyparissias* L.). *Biol. Cont.*, **5**, 466–477.

Goeden, R.D. (1993) Arthopods for the suppression of terrestrial weeds. *Proc. Beltsville Symposium 18, Pest Management: Biologically Based Technologies*, ARS, USDA, American Chemical Society, Washington DC, pp. 231–237.

Goeden, R.D. and Palmer, W.A. (1995) Lessons learned from studies of the insects associated with Ambrosiinae in North America in relation to the biological control of weedy members of this group. In E.S. Delfosse and R.R. Scott (eds.), *Proc. 8th Int. Symp. Biol. Control Weeds*, CSIRO, Melbourne, pp. 565–573.

Grodowitz, M.J., Jeffers, L., Graham, S. and Nelson, M. (1996) Innovative approaches to transferring information on the use of biological control for noxious and nuisance plant management. In V.C. Moran and J.H. Hoffmann (eds.), *Proc. 9th Int. Symp. Biol. Control Weeds*, University of Capetown, Stellenbosch, pp. 269–272.

Harley, K.L.S. and Forno, I.W. (1992) *Biological Control of Weeds: A Handbook for Practitioners and Students*, Inkata, Melbourne, Australia, 74 pp.

Harley, K.L.S., Kassulke, R.C., Sands, D.P.A. and Day, M.D. (1990) Biological control of water lettuce, *Pistia stratiotes* (Araceae), by *Neohydronomus affinis* (Coleoptera: Curculionidae). *Entomophaga*, **35**, 363–374.

Harley, K., Gillett, J., Winder, J., Forno, W. and Segura, R., *et al.* (1995) Natural enemies of *Mimosa pigra* and *M. berlandieri* (Mimosaceae) and prospects for biological control of *M. pigra. Environ. Entomol.* **24**, 1664–1678.

Harris, P. (1984) *Carduus nutans* L., nodding thistle, and *C. acanthoides* L., plumeless thistle (Compositae). In J.S. Kelleher and M.A. Hulme (eds.), *Biological Control Programmes against Insects and Weeds in Canada 1969–1980*, CAB, Farnham Royal, pp. 115–126.

Harris, P. (1991) Classical biocontrol of weeds: its definition, selection of effective agents, and administrative-political problems. *Can. Entomol.*, **123**, 827–849.

Harris, P. (1993) Effects, constraints and the future of weed biocontrol. *Agric. Ecosyst. Environ.*, **46**, 289–303.

Harris, P. and McEvoy, P. (1995) The predictability of insect host plant utilization from feeding tests and suggested improvements for screening weed biological control agents. In E.S. Delfosse and R.R. Scott (eds.), *Proc. 8th Int. Symp. Biol. Control Weeds*, CSIRO, Melbourne, pp. 125–131.

Hight, S.D., Blossey, B., Laing, J. and DeClerke-Float, R. (1995) Establishment of insect biological control agents from Europe against *Lythrum salicaria* in North America. *Environ. Entomol.*, **24**, 967–977.

Hirose, Y. (1992) *Biological Control in South and East Asia*, Fukuoka, Japan, IOBC/SEARS and Kyushu University Press, 68 pp.

Hoffmann, J.H. (1991) Biological control of weeds in South Africa: Special Issue. *Agric. Ecosyst. Environ.*, **37**, 1–3.

Hoffmann, J.H. (1995) Biological control of weeds: the way forward, a South African perspective. *Proc. BCPC Symp. Weeds in a Changing World*, **64**, 77–89.

Hoffmann, J.H. and Moran, V.C. (1995) Biological control of *Sesbania punicea* with *Neodiplogrammus quadrivittatus*: predictions of limited success soon confounded. In E.S. Delfosse and R.R. Scott (eds.), *Proc. 8th Int. Symp. Biol. Control Weeds*, CSIRO, Melbourne, p. 201.

Hoffmann, J.H. and Moran, V.C. (1998) The population dynamics of an introduced tree, *Sesbania punicea*, in South Africa, in response to long-term damage caused by different combinations of three species of biological control agents. *Oecologia*, **114**, 343–348.

Holden, A.N.G. and Mahlberg, P.G. (1995) Rusts for the biological control of leafy spurge (*Euphorbia esula*) in North America. In E.S. Delfosse and R.R. Scott (eds.), *Proc. 8th Int. Symp. Biol. Control Weeds*, CSIRO, Melbourne, pp. 419–424.

Holtkamp, R.H. and Campbell, M.H. (1995) Biological control of *Cassinia* spp. (Asteraceae). In E.S. Delfosse and R.R. Scott (eds.), *Proc. 8th Int. Symp. Biol. Control Weeds*, CSIRO, Melbourne, pp. 447–450.

Hosking, J.R., McFadyen, R.E.C. and Murray, N.D. (1988) Distribution and biological control of cactus species in eastern Australia. *Plant Prot. Quart.*, **3**, 115–123.

Howarth, F.G. (1991) Environmental impacts of classical biological control. *Annu. Rev. Entomol.*, **36**, 485–509.

Hughes, C.E. (1995) Protocols for plant introductions with particular reference to forestry: changing perspectives on risks to biodiversity and economic development. *Proc. BCPC Symp. Weeds in a Changing World*, **64**, 15–32.

Humphries, S.E., Groves, R.H. and Mitchell, D.S. (1991) Plant invasions of Australian ecosystems. *Kowari*, **2**, 1–134.

Jayanth, K.P. and Ganga Visalakshy, P.N. (1996) Succession of vegetation after suppression of parthenium weed by *Zygogramma bicolorata* in Bangalore, India. *Biol. Agric. Hort.*, **12**, 303–309.

Jobin, A., Schaffner, U. and Nentwig, W. (1996) The structure of the phytophagous insect fauna on the introduced weed *Solidago altissima* in Switzerland. *Entomol. Exp. Appl.*, **79**, 33–42.

Johnson, D.M. and Stiling, P.D. (1996) Host specificity of *Cactoblastis cactorum* (Lepidoptera: Pyralidae), an exotic *Opuntia*-feeding moth, in Florida. *Environ. Entomol.*, **25**, 743–748.

Julien, M.H., and Chan, R.R. (1992) Biological control of alligator weed: unsuccessful attempts to control terrestrial growth using the flea beetle *Disonycha argentinensis* (Coleoptera: Chrysomelidae). *Entomophaga*, **37**, 215–221.

Julien, M.H. and Griffiths, M. (1998) *Biological Control of Weeds: A World Catalogue of Agents and their Target Weeds*, CAB International, Wallingford, Oxford.

Julien, M.H., Harley, K.L.S., Wright, A.D., Cilliers, C.J., Hill, M.P., *et al.* (1996) International cooperation and linkages in the management of water hyacinth with emphasis on biological control. In V.C. Moran and J.H. Hoffmann (eds.), *Proc. 9th Int. Symp. Biol. Control Weeds*, University of Capetown, Stellenbosch, pp. 273–282.

Jupp, P.W. and Cullen, J.M. (1996) Expected and observed effects of the mite *Aculus hyperici* on St John's wort, *Hypericum perforatum*, in Australia. In V.C. Moran and J.H. Hoffmann (eds.), *Proc. 9th Int. Symp. Biol. Control Weeds*, University of Capetown, Stellenbosch, pp. 365–370.

Kelly, D. and McCallum, K. (1995) Evaluating the impact of *Rhinocyllus conicus* on *Carduus nutans* in New Zealand. In V.C. Moran and J.H. Hoffmann (eds.), *Proc. 9th Int. Symp. Biol. Control Weeds*, University of Capetown, Stellenbosch, pp. 205–211.

Kroschel, J. and Klein, O. (1999) Biological control of *Orobanche* spp. with *Phytomyza orobanchia* Kalt., a review. In J. Kroschel, M. Abderabihi, and H. Betz (eds.), *Advances in Parasitic Weed Control at On-farm Level*, vol. II, *Joint Action to Control* Orobanche *in the WANA Region*, Margraf-Verlag, Weikersheim, pp. 135–159.

Kuniata, L.S. (1994) Importation and establishment of *Heteropsylla spinulosa* (Homoptera: Psyllidae) for the biological control of *Mimosa invisa* in Papua New Guinea. *Int. J. Pest Manage.*, **40**, 64–65.

Labrada, R. (1996) The importance of biological control for the reduction of the incidence of major weeds in developing countries. In V.C. Moran and J.H. Hoffmann (eds.), *Proc. 9th Int. Symp. Biol. Control Weeds*, University of Capetown, Stellenbosch, pp. 287–290.

Lanaud, C., Jolivot, M.P. and Deat, M. (1991) Preliminary results on the enzymatic diversity in *Chromolaena odorata* (L.) In R.M. King and H. Robinson (eds.), *Proc. Int. Workshop on Biological Control and Management of Chromolaena odorata*, Bogor, pp. 71–77.

Lonsdale, W.M. (1994) Inviting trouble: introduced pasture species in northern Australia. *Austr. J. Ecol.*, **19**, 345–354.

Lonsdale, W.M., Miller, I.L. and Forno, I.W. (1989) The biology of Australian weeds 20. *Mimosa pigra* L. *Plant Prot. Quart.*, **4**, 119–131.

Markin, G.P., Lai, P-Y. and Funusaki, G.P. (1992) Status of biological control of weeds in Hawaii and implications for managing native ecosystems. In C.P. Stone, C.W. Smith and J.T. Tunison (eds.), *Alien Plant Invasions in Native Ecosystems of Hawaii: Management and Research*, University of Hawaii Press, Honolulu, pp. 466–482.

Marohasy, J. (1996) Host shifts in biological weed control: real problems, semantic difficulties or poor science? *Int. J. Pest Manage.*, **42**, 71–75.

Marsden, J.S., Martin, G.E., Parham, D.J., Risdill-Smith, T.J. and Johnston, B.G. (1980) Skeleton weed control. In *Returns on Australian Agricultural Research*, CSIRO Division of Entomology, Canberra, pp. 84–93.

McClay, A.S. (1995) Beyond "before-and-after": experimental design and evaluation in classical weed biological control. In E.S. Delfosse and R.R. Scott (eds.), *Proc. 8th Int. Symp. Biol. Control Weeds*, CSIRO, Melbourne, pp. 213–219.

McClay, A.S. (1996a) Host range, specificity and recruitment: synthesis of session 2. In V.C. Moran and J.H. Hoffmann (eds.), *Proc. 9th Int. Symp. Biol. Control Weeds*, University of Capetown, Stellenbosch, pp. 105–112.

McClay, A.S. (1996b) Biological control in a cold climate: temperature responses and climatic adaptation of weed biocontrol agents. In V.C. Moran and J.H. Hoffmann (eds.), *Proc. 9th Int. Symp. Biol. Control Weeds*, University of Capetown, Stellenbosch, pp. 377–383.

McClay, A.S., Palmer, W.A., Bennett, F.D. and Pullen, K.R. (1995) Phytophagous arthropods associated with *Parthenium hysterophorus* (Asteraceae) in North America. *Environ. Entomol.*, **24**, 796–809.

McEvoy, P.B., Cox, C.S. and Coombs, E.M. (1991) Successful biological control of ragwort. *Ecol. Applic.*, **1**, 430–432.

McFadyen, P.J. (1985) Introduction of the gall fly *Rhopalomyia californica* from the USA into Australia for the control of the weed *Baccharis halimifolia*. In E.S. Delfosse (ed.), *Proc. 6th Int. Symp. Biol. Control Weeds*, Agriculture Canada, Ottawa, pp. 779–787.

McFadyen, R.E.C. (1986) *Harrisia* cactus in Queensland. In R. Kitching (ed.), *The Ecology of Exotic Animals and Plants in Australia*, Jacaranda-Wiley, Brisbane, pp. 240–261.

McFadyen, R.E.C. (1991) Climate modelling and the biological control of weeds: one view. *Plant Prot. Quart.*, **6**, 14–15.

McFadyen, R.E.C. (1998) Biological control of weeds. *Annu. Rev. Entomol.*, **43**, 369–393.

McFadyen, R.E.C. (2000) Successes in biological control. In N.R. Spencer (ed.), *Proc. Xth. Int. Symp. Biol. Control Weeds*, Montana State University, Bozeman, MT, pp. 3–14.

McFadyen, R.E.C. (2001) Recent developments in biocontrol of weeds using insects. In S.P. Singh, B.S. Bhumannavar, J. Poorani and D. Singh (eds.), *Biological Control*, Project Directorate of Biological Control, Bangalore, and Punjab Agricultural University, Ludhiana, India, pp. 157–162.

McFadyen, R.E.C. (2002) Impacts of exotic weeds on wildlife. In A.J. Franks, J. Playford and A. Shapcott (eds.), *Landscape Health of Queensland*, The Royal Society of Queensland, St. Lucia, Queensland, pp. 128–132.

McFadyen, R.E.C. and McClay, A.S. (1981) Two new insects for the biological control of parthenium weed in Queensland. *Proc. Austr. Weeds Conf., 6th, Gold Coast*, Qld. Weed Soc, Brisbane, pp. 145–149.

McFadyen, R.E.C. and Marohasy, J.J. (1990a) A leaf-feeding moth, *Euclasta whalleyi* (Lepidoptera: Pyralidae), for the biological control of *Cryptostegia grandiflora* (Asclepiadaceae) in Queensland, Australia. *Entomophaga*, **35**, 431–435.

McFadyen, R.E.C. and Marohasy, J.J. (1990b) Biology and host plant range of the soft scale *Steatococcus* new species (Hem.: Margarodidae) for the biological control of the weed *Cryptostegia grandiflora* (Asclepiadaceae). *Entomophaga*, **35**, 437–439.

Moran, V.C. and Zimmermann, H.G. (1991) Biological control of jointed cactus, *Opuntia aurantiaca* (Cactaceae) in South Africa. *Agric. Ecosyst. Environ.*, **37**, 5–27.

Morin, L. (1996) Different countries, several potential bioherbicides, but always the same hurdles. In V.C. Moran and J.H. Hoffmann (eds.), *Proc. 9th Int. Symp. Biol. Control Weeds*, University of Capetown, Stellenbosch, p. 546 (Abstr.).

Morin, L., Auld, B.A. and Smith, H.E. (1996) Rust epidemics, climate and control of *Xanthium occidentale*. In V.C. Moran and J.H. Hoffmann (eds.), *Proc. 9th Int. Symp. Biol. Control Weeds*, University of Capetown, Stellenbosch, pp. 385–391.

Morris, M.J. (1997) Impact of the gall-forming rust fungus *Uromycladium tepperianum* on the invasive tree *Acacia saligna* in South Africa. *Biol. Cont.*, **10**, 75–82.

Nissen, S.J., Masters, R.A., Lee, D.J. and Rowe, M.L. (1995) DNA-based marker systems to determine genetic diversity of weedy species and their application to biocontrol. *Weed Sci.*, **43**, 504–513.

Nordlund, D.A. (1996) Biological control, integrated pest management and conceptual models. *Biocont. News Inf.*, **17**, 35–44.

Olckers, T., Zimmermann, H.G. and Hoffmann, J.H. (1995) Interpreting ambiguous results of host-specificity tests in biological control of weeds: assessment of two *Leptinotarsa* species (Chrysomelidae) for the control of *Solanum elaeagnifolium* (Solanaceae) in South Africa. *Biol. Cont.*, **5**, 336–344.

Ooi, P.A.C. (1987) A fortuitous biological control of *Lantana* in Malaysia. *PANS*, **33**, 234–235.

Ooi, P.A.C. (1992) Biological control of weeds in Malaysian plantations. *Proc. Int. Weed Cont. Congr.*, Weed Science Society Victoria, Melbourne, pp. 248–255.

Palmer, W.A. and Goeden, R.D. (1991) The host range of *Ophraella communa* LeSage (Coleoptera: Chrysomelidae). *Coleopt. Bull.*, **45**, 115–120.

Palmer, W.A. and Miller, E.N. (1996) A method for prioritising biological control projects with reference to those of Queensland. In V.C. Moran and J.H. Hoffmann (eds.), *Proc. 9th Int. Symp. Biol. Control Weeds*, University of Capetown, Stellenbosch, pp. 313–317.

Panetta, F.D. (1993) A system of assessing proposed plant introductions for weed potential. *Plant Prot. Quart.*, **8**, 10–14.

Paton, R. (1995) Legislation and its administration in the approval of agents for biological control in Australia. In E.S. Delfosse and R.R. Scott (eds.), *Proc. 8th Int. Symp. Biol. Control Weeds*, CSIRO, Melbourne, pp. 653–658.

Pemberton, R.W. (1995) *Cactoblastis cactorum* (Lepidoptera: Pyralidae) in the United States: an immigrant biological control agent or an introduction of the nursery industry? *Am. Entomol.*, **41**, 230–232.

Peschken, D.P. and McClay, A.S. (1995) Picking the target: a revision of McClay's scoring system to determine the suitability of a weed for classical biological control. In E.S. Delfosse and R.R. Scott (eds.), *Proc. 8th Int. Symp. Biol. Control Weeds*, CSIRO, Melbourne, pp. 137–143.

Pimentel, D. (1995) Biotechnology: environmental impacts of introducing crops and biocontrol agents in North American agriculture. In H.M.T. Hokkanen and J.M. Lynch (eds.), *Biological Control: Benefits and Risks*, Cambridge University Press, Cambridge, UK, pp. 13–29.

Pomerinke, M.A., Thompson, D.C. and Clason, D.L. (1995) Bionomics of *Cleonidius trivittatus* (Coleoptera: Curculionidae): native biological control of purple locoweed (Rosales: Fabaceae). *Environ. Entomol.*, **24**, 1696–1702.

Popay, K. and Field, R. (1996) Grazing animals as weed control agents. *Weed Technol.*, **10**, 217–231.

Radford, I.J., Muller, P., Fiffer, S. and Michael, P.W. (2000) Genetic relationships between Australian fireweed and South African and Madagascan populations of *Senecio madagascariensis* Poir. and closely related *Senecio* species. *Aust. Syst. Bot.*, **13**, 409–423.

Randall, J.M. (1996) Weed control for the preservation of biological diversity. *Weed Technol.*, **10**, 370–383.

Rayachhetry, M.B., Blakeslee, G.M. and Center, T.D. (1996) Predisposition of Melaleuca (*Melaleuca quinquenervia*) to invasion by the potential biological control agent *Botryosphaeria ribis*. *Weed Sci.*, **44**, 603–608.

Reichard, S.E. (1997) Prevention of invasive plant introductions on national and local levels. In J.O. Luken and J.W. Thieret (eds.), *Assessment and Management of Plant Invasions*, Springer, New York, pp. 215–227.

Reznik, S.Y. (1996) Classical biocontrol of weeds in crop rotation: a story of failure and prospects for success. In V.C. Moran and J.F. Hoffmann (eds.), *Proc. 9th Int. Symp. Biol. Control Weeds*, University of Capetown, Stellenbosch, pp. 503–506.

Richardson, D.M. and Higgins, S.I. (1997) Pines as invaders in the southern hemisphere. In D.M. Richardson (ed.), *Ecology and Biogeography of Pinus*, Cambridge University Press, Cambridge, UK.

Sands, D.P.A. and Schotz, M. (1985) Control or no control: a comparison of the feeding strategies of two salvinia weevils. In E.S. Delfosse (ed.), *Proc. 6th Int. Symp. Biol. Control Weeds*, Agriculture Canada, Ottawa, Canada, pp. 551–556.

Scheepens, P.C., Muller-Scharer, H. and Kempenaar, C. (2001) Opportunities for biological weed control in Europe. *BioControl*, **46**, 127–138.

Schroeder, D. and Goeden, R.D. (1986) The search for arthropod natural enemies of introduced weeds for biological control — in theory and practice. *Biocont. News Inf.*, **7**, 147–155.

Scott, J.K. (1996) Population ecology of *Chrysanthemoides monilifera* in South Africa: implications for its control in Australia. *J. Appl. Ecol.*, **33**, 1496–1508.

Shea, K. and Possingham, H.P. (2000) An application of stochastic dynamic programming to population management. *J. Appl. Ecol.*, **37**, 77–86.

Sheppard, A.W., Briese, D.T. and Michalakis, I. (1995) Host choice in the field in the genus *Larinus* (Coleoptera: Curculionidae) attacking *Onopordum* and *Cynara* (Asteraceae). In E.S. Delfosse and R.R. Scott (eds.), *Proc. 8th Int. Symp. Biol. Control Weeds*, CSIRO, Melbourne, pp. 605–614.

Siebert, T. (1989) Biological control of the weed *Chromolaena odorata* (Asteraceae) by *Pareuchaetes pseudoinsulata* (Lep.: Arctiidae) on Guam and the Northern Mariana Islands. *Entomophaga*, **34**, 531–539.

Simberloff, D. and Stiling, P. (1996) How risky is biological control? *Ecology*, **77**, 1965–1974.

Simmonds, F.J. and Bennett, F.D. (1966) Biological control of *Opuntia* spp. by *Cactoblastis cactorum* in the Leeward Islands (West Indies). *Entomophaga*, **11**, 183–189.

Spollen, K.M. and Piper, G.L. (1995) Effectiveness of the gall mite, *Eriophyes chondrillae*, as a biological control agent of rush skeletonweed (*Chondrilla juncea*) seedlings. In E.S. Delfosse and R.R. Scott (eds.), *Proc. 8th Int. Symp. Biol. Control Weeds*, CSIRO, Melbourne, pp. 375–379.

Syrett, P. (1995) An environmental impact assessment for biological control of heather (*Calluna vulgaris*) in New Zealand. In V.C. Moran and J.H. Hoffmann (eds.), *Proc. 9th Int. Symp. Biol. Control Weeds*, University of Capetown, Stellenbosch, pp. 69–74.

Timbilla, J.A. and Braimah, H. (2002) Successful biological control of *Chromolaena odorata* in Ghana: potential for a regional programme in Africa. In: C. Zachariades, R. Muniappan and L.W. Strathie (eds.), *Proc. Int. Workshop on Biological Control and Management of Chromolaena odorata*, Durban (in press).

Thomas, P.A. and Room, P.M. (1985) Towards biological control of salvinia in Papua New Guinea. In E.S. Delfosse (ed.), *Proc. 6th Int. Symp. Biol. Control Weeds*, Agriculture Canada, Ottawa, Canada, pp. 567–574.

Turner, C.E., Pemberton, R.W. and Rosenthal, S.S. (1987) Host utilization of native *Cirsium* thistles (Asteraceae) by the introduced weevil *Rhinocyllus conicus* (Coleoptera: Curculionidae) in California. *Environ. Entomol.*, **16**, 111–115.

Vitelli, M., James, P., Marohasy, J., McFadyen, R.E., Trevino, M. and Hannan-Jones, M. (1996) Field establishment — how long does it take? *Proc. 9th Int. Symp. Biol. Control Weeds*, University of Capetown, Stellenbosch, p. 421 (Abstr.).

Wan, F-H. and Harris, P. (1996) Host finding and recognition by *Altica carduorum*, a defoliator of *Cirsium arvense*. *Entomol. Exp. Appl.*, **80**, 491–496.

Wapshere, A.J. (1989) A testing sequence for reducing rejection of potential biological control agents for weeds. *Ann. Appl. Biol.*, **114**, 515–526.

Wapshere, A.J., Delfosse, E.S. and Cullen, J.M. (1989) Recent developments in biological control of weeds. *Crop Prot.*, **8**, 227–250.

Webb, M. and Conroy, C. (1995) The socio-economics of weed control on smallholder farms in Uganda. *Proc. Brighton Crop Prot. Conf. — Weeds*, **1**, 157–162.

Williamson, M. and Fitter, A. (1996) The varying success of invaders. *Ecology*, **77**, 1661–1668.

Woodburn, A.T. (1995) The market for agrochemicals: present and future. *Proc. Brighton Crop Prot. Conf. — Weeds*, **1**, 121–128.

Woodburn, T.L. (1996) Interspecific competition between *Rhinocyllus conicus* and *Urophora solstitialis*, two biocontrol agents released in Australia against *Carduus nutans*. In V.C. Moran and J.H. Hoffmann (eds.), *Proc. 9th Int. Symp. Biol. Control Weeds*, University of Capetown, Stellenbosch, pp. 409–415.

Species Index

Acacia saligna 173
Acrobasis vaccinii 65
Agathis diversa 43
Ageratina riparia 173
Alternanthera phileroxoides 173
Amblyseius californicus 43
Amblyseius potentillae 94
Ametacloria misella 147
Amorbia cuneana 45
Anagrus epos 25
Anaphes iole 43, 48
Anthonomus grandis 18, 22, 25, 26
Aonidiella aurantii 9, 30, 131
Apanteles glomeratus 146
Apanteles medicaginis 52
Apanteles melanoscelus 130, 146
Aphidius smithi 43
Aphis gossypii 48
Aphytis chrysomphali 30, 131
Aphytis lingnanensis 9, 30, 54, 131
Aphytis melinus 30, 43, 131
Archytas marmoratus 43, 48, 54
Aschesonia aleyrodis 156
Asobara persimillis 126
Asobara rufescens 126
Asobara spec 126
Asobara tabida 126
Astralagus mallissimus 164

Bacillus popilliae 145
Bacillus thuringiensis 25, 32, 75, 146, 147
Bacillus thuringiensis japonensis 145
Bacillus thuringiensis kurstaki 45
Bactrocera cucurbitae 123
Bactrocera dorsalis 130
Bathyplectes anurus 22
Bathyplectes curculionis 22, 26
Beauveria bassiana 145
Bemisia argentifolii 31, 43, 47, 54
Bemisia tabaci 23, 54
Binckochrysa scelertes 12
Bousseola fusca 65
Brachymeria lasus 55
Bracon kirkpatricki 54
Brassica kaber 24
Brassica vulgaris 24
Brevicoryne brassicae 114
Bruchus pisorum 24

Cactobastis cactorum 174
Calluna vulgaris 168
Campoletis sonorensis 24, 25, 55, 124
Cardiochiles nigriceps 127
Carduus nutans 171, 173
Catolaccus grandis 25–29, 31, 49, 50, 53–55
Ceratitis capitata 22, 131
Ceutorhynchus assimilis 26
Chilo infuscatellus 54
Chilo partellus orichalociliellus 65
Chilo sacchariphagus 54
Choristoneura rosaceana 45
Chromolaena odorata 173, 176
Chrysanthemoides 170
Chrysopa carnea 43, 54
Chrysoperla carnea 9, 25, 29, 30, 48, 55, 151
Chrysoperla rufilabris 29
Cleonidius trivittatus 164
Coccinella septempunctata 43, 115, 148, 151
Coleomegilla maculata 25
Colias eurytheme 52
Compsilura concinnata 130
Cotesia kariyani 9
Cotesia marginiventris 55, 90
Cotesia melanoscela 48, 54
Cotesia plutellae 26
Corcyra cephalonica 12
Cordia curassavica 173
Cryptolaemus montronzieri 43, 54
Cryptostegia grandiflora 167
Cynoglossum 170
Cyrtobagous salviniae 165, 173
Cyrtobagous singularis 173

Delphastus pusillus 43, 47, 54
Dendranthema grandiflora 48
Dendrolimus punctata 54
Diachasmimorpha longicaudata 123, 130
Diadegma insularis 24
Diadegma semiclausum 26, 147
Diaeretiella rapae 43
Diapetimorpha introita 28
Diatraea saacharalis 71, 74
Diglyphus begini 43, 47, 54
Dysdercus superstitiosus 95

Echium plantagineum 166
Edouum puttleri 24, 32
Eichhornia crassipes 173
Eladana saccharina 65
Encarsia formosa 26, 30, 43, 47, 48, 54, 156
Encarsia luteola 43, 47, 54
Ephestia elutella 66
Ephestia kuehniella 66, 69, 70
Ephestia sericarium 127
Epidinocarsis diversicornis 130
Epidinocarsis lopezi 4, 30, 130
Epilachna varivestis 22, 89, 93
Epiphyas postvittana 45, 54, 67, 70, 75
Epiricania melanoleuca 8
Episysphus balteatus 151
Eretmocerus californicus 43, 47
Eretmocerus eremicus 30, 47, 54
Erynia neoaphidis 148, 151
Erythroneura elegantula 25, 48
Erythroneura variabilisi 48
Eucelatoria bryani 55, 129
Eucelatoria rubentis 129
Eupteromalus leguminis 24
Euschistus conspersus 129
Euseius tularensis 43, 48, 54
Exeristes roborator 55

Fopius arisanus 130
Fusarium solani 95

Geocoris punctipes 29
Glycine max 89, 91, 92
Glypta fumiferanae 128

Hamakuapar pamakani 173
Harrisia cactus 173
Harrisia martinii 173
Harrisina brillians 147
Hedychium gardnerianum 167
Helianthus annuus 169
Helianthus ciliaris 169
Helicoverpa armigera 12, 54, 66, 72, 74
Helicoverpa zea 11, 23, 26, 43, 45, 46, 48,
    54, 129, 145
Heliothis virescens 23–25, 46, 127
Hippodamia convergens 25, 43, 48, 151
Horogenus choysostictos 127, 128
Hypera postica 22, 26
Hypericum perforatum 173

Icerya purchasi 3, 22
Itoplectis conquisitor 55

Jalysus spinosus 43
Jasamina calamistis 65

Leptinotarsa decemlineata 32
Leptomastix dactylopii 43
Liriomyza trifolii 43, 47, 54
Lixophaga diatraeae 43, 54, 55,
    128
Lygus hesperus 43, 48
Lygus lineolaris 46
Lysiphlebus testaceipes 43, 54

Macquartia chalconota 127
Macrocentrus ancylivorus 43, 54
Mamestra brassicae 45
Metaphycus helvolus 54
Metaphycus lounsburyi 131, 132
Metasciulus occidentalis 9
Microchelonus blackburni 54
Microctonus aethiopoides 22
Microplitis croceips 53, 55, 90, 145
Microplitis demolitor 90–94
Microterys flavus 43
Mimosa invisa 173
Mimosa pigra 167, 170
Monomorium pharaonis 3
Musca domestica 131
Muscidifurax raptor 31
Muscidifurax raptorellus 31
Muscidifurax zaraptor 31
Myiopharus doryphorae 146
Mythimna separata 9
Myzus persicae 48, 116

Neosciulus fallacis 48, 54
Nemeritis canescens 127, 128

Oecophylla smaragdina 3
Ooencyrtus kuvanae 124
Ooencyrtus nezarae 71
Oomyzus incertus 22, 30
Ophraella communa 169
Opius humilis 131
Opius tryoni 131
Opuntia triacantha 174
Ostrinia nubilalis 25, 66, 72, 74

Pachycrepoideus vinclemia 55
Parlatoria olrae 22
Passiflora mallissima 172
Patanga septemfasciata 3
Pectinophora gossipiella 74
Pediobius foveolatus 22, 30, 44, 54
Perillus bioculatus 25, 29, 32, 146
Periplaneta americana 96
Phacelia tanacetifolia 24
Phaseolus limensis 54
Phaseolus lunatus 54

Phenacoccus manihoti 4, 130
Phyllotreta cruciferae 114
Phytomyza orobanchia 164
Phytoseiulus persimilis 43, 94
Pieris brassicae 146
Pimpla turionella 152
Pistia stratiotes 173
Plutella xylostella 24, 26, 45, 72
Podisus maculoventris 29, 151
Podisus sagitta 29
Pseudoplusia includens 91–94
Psidium cattleianum 167
Psix tumetanus 129
Psyttalia fletcheri 123
Psyttalia incisi 130
Pteromalus puparum 55
Pterostichus melanarius 115
Puccinia xanthii 171

Ratibida pinnata 170
Rhinocyllus conicus 174
Rhopalomyia californica 131
Rhopobota naevana 65
Rhyacionia buoliana 24
Rodolia cardinalis 3, 22
Rogas lymantriae 146

Sabulodes aegrotata 45
Saissetia oleae 131, 132
Salvinia molesta 164, 173
Sarcophaga shutzei 145
Scirtothrips citri 48, 54
Scolytus multistriatus 96
Scutellista cyanea 131, 132
Senecio jacobaeae 172
Serbania punicea 173
Sitotroga cerelalella 54, 69, 70, 75
Solenopsis invicta 27
Solidago altrisima 166
Spalangia cameroni 31
Spalangia endius 31
Spalangia nigroaenea 31
Spodoptera exigua 23
Spodoptera frugiperda 25

Spodoptera praefica 132
Stethorus picipes 44
Streptomyces avermitilis 26

Telenomus heliothidis 125
Tetranychus urticae 26, 48, 94
Theobroma cacao 96
Therioaphis trifolii 22
Trialeurodes vaporariorum 26, 30, 47, 48
Trichogramma brassicae 68, 70–74, 76
Trichogramma cacoecia 44
Trichogramma carverae 44, 45, 54, 70, 71, 76, 79, 80
Trichogramma chilonis 8, 9, 12, 72
Trichogramma confusum 54
Trichogramma dendrolini 54, 55
Trichogramma embryophagum 66
Trichogramma evanescens 66, 124
Trichogramma fumiculatum 76
Trichogramma galloi 71, 74
Trichogramma ivelae 76
Trichogramma maidis 68
Trichogramma minutum 44, 65, 70
Trichogramma nubiale 44
Trichogramma platneri 45
Trichogramma pretiosum 11, 44–46, 55, 70, 72, 74, 125
Trichogramma sibericum 65
Trichogrammatoidea bactrae 76
Trichogrammatoidea nana 8
Trichomalus perfectus 26
Trichoplusia ni 45, 90, 93, 132
Trissolcus basalis 129
Trissolcus utahensis 129
Tyria jacobaeae 174

Veneza phyllopus 93

Xanthium occidentale 171
Xyloborus ferrugineus 95, 96, 104

Zoophthora phytonomi 22
Zoophthora radicans 147
Zygogramma bicolorata 174

Subject Index

Abamectin 26, 47
Additive 142
Additive effect 141
Adelphoparasitoids 7
Agricultural pests 18
Agroecosystems 8–11, 18, 24
Alate aphids 113
Aldicarb 26
Alfalfa weevils 22, 26, 30
Angoumois grain moth 54, 75
Antagonism 140
Aphid parasites 43
Aphids 7, 24, 44, 48, 115, 116, 151
Applied biological control 1
Artificial diets 28, 55
Assassin bugs 5
Augmentation 8, 21, 27, 41, 42, 164
Augmentative biocontrol 9, 41
Avocado brown mite 44

Bt crops 117
Baculoviruses 15
Beet armyworm 23, 54
Beetles 5
Bioherbicides 164
Biological control 1, 2, 4, 7, 9, 18, 19, 121, 154, 163, 164
Blackheaded fireworm 65
Black scale 54
Boll weevil 18, 23, 25–28, 31, 49, 50, 54, 55
Bollworm 44
Braconid endoparasitoid 22
Braconid parasites 141, 151
Braconid wasp 12
Brown soft scale 43

Cabbage aphids 114
Cabbage looper 44, 45, 93, 132, 151
California red scale 9, 30, 43
Carabid beetle 115
Carbaryl-OP-sulfur- OP resistant strain 9
Cassava mealybug 4, 130
Chemical ecology 93
Chemoanemotactic orientation 11
Chlorpyrifos 73
Cinnabar moth 174
Citrus mealy bug 43, 54

Citrus thrips 43
Classical biocontrol 164, 165
Coccinellid beetle 47, 151
Codling moth 44
Coleoptera 5
Colorado potato beetle 24, 25, 29, 32, 146, 147
Conservation 9, 21, 23, 164
Cotton bollworm 8
Cotton fleahoppers 49
Cotton stainer bug 95
Cottony cushion scale 3, 7, 22
Cranberry fruitworm 65
Cultural practices 23
Cyclamen mite 44

DDT 22
Decosane 12
Deterrent 115
Diamondback moth 8, 24, 26
Dimethoate 46
Dipel 45
Diptera 5, 6
Discrimination of parasitized hosts 125
Diversity and natural enemies 115
Dragon flies 5

Ecological selectivity 26
Egg parasitoids 12, 48, 65, 74
Endogram 9
Endoparasitoids 55
Endosulfan tolerant strain 9
δ-Endotoxin 45
Entomophage park 3
Entomophagous insects 9, 139
Environmental energy exchange code 98
Epizootics 150
Eulophid 24; larval endoparasitoid 22
European corn borer 25, 44, 45, 66
Exotic insects 163
Extrinsic chemical defenses 88, 90

Facultative pathogen 145
Fall armyworm 25
Filth flies 44, 54
Flea beetles 114
Fly parasites 44, 54

Foraging parasitoid 147
Fungal pathogen 22

Glucosinolates 115
Granulovirus 146
Greater wax moth 54
Green bug 43, 54
Greenhouse white fly 26, 30, 43, 54
Green lacewings 5
Green peach aphid 43
Ground beetles 116
Gypsy moth 48, 54, 130, 145, 146

Herbivores 1
Hexatriacontane 12
High quality wasps 77
Host parasitoid synchrony 72
Host plant 18
Host preference 9
Host range 9
Host recognition 124
Host response behavior 123
Housefly 54, 131
Hover fly 5
Hymenoptera 5, 6
Hyperparasites 152
Hyperparasitism 2, 121
Hyperparasitoids 2, 7

Ichneumonid parasites 151
Idiobiontic parasitoids 29
Importation 19, 22
Imported cabbageworm 44
Independent additive 142
Inducible plant defenses 87, 96, 97, 98
Inoculative release 8, 29, 66
Insect pathogens 139
Integrated pest management 12, 13, 17, 18, 23, 24, 65, 121
Integration issues 78
Intercropping 24
International organization of biological Control 7
Intrinsic chemical defenses 88, 89
Inundative release 8, 28, 30, 31

Kairomones 12, 46, 87, 90
Koinobiontic parasitoids 29

Lacewings 49, 151
Ladybird beetles 115, 116, 151
Lantana 165
Larval ectoparasitoids 55
Leaf footed bug 93
Leafhopper 48, 49

Leafroller 44
Lightbrown apple moth 67, 75–77

Malathion 22, 23, 26
Mass propagation 29
Mass release 48
McDaniel spider mite 44
Mealy bugs 43, 164
Mechanisms of competition 126
Mediterranean black scale 131, 132
Mediterranean fruit fly 22
Melon aphids 43, 48
Melon fly larva 123
Meridic diets 28, 29
Methomyl 26, 45
Methyl parathion 26
Mexican bean beetle 22, 30, 44, 54, 89, 90, 93
Microhabitat 11
Monitoring of natural enemies 52
Monocosane 12
Monocrotophos tolerant strain 9
Monophagous 5
Multiparasitism 121; competitive displacement 130; costs 129; physical struggle 127; physiological inhibition 128
Multiple cropping 24
Muscoid flies 31
Mycangia 95
Mycoherbicides 164

Natural enemies 1, 7, 8, 11, 12, 17, 21, 22, 25, 41, 50, 111, 143
Natural enemy conservation 26, 27
Neuroptera 5
Nodding thistle 173
Non-additive 142
Nonclassical biocontrol 164
Nonintervention decision tactic 27
Noogoora burr 171–173
NPV 145, 151

Obliquebanded leafroller 45
Odonata 5
Oleander scale 54
Oligophagous 5
Olive scale 22
Omnivorous looper 45
Organophosphates 23
Oriental fruit fly 129
Oriental fruit moth 43, 54

Parasites 6, 54, 146, 155
Parasitic mites 5
Parasitic nematodes 145
Parasitization 11

Parasitization behavior 121
Parasitoid competition 122
Parasitoid-host relationship 9
Parasitoid wasp 71
Parasitoids 1, 6, 24, 146
Pathogens 155
Patterson's curse 166
Pea aphid 43, 54, 148
Pea weevils 24
Pentatomid predators 151
Pentatomid stink bug 129
Pest management 18
Pesticide tolerant strain 9
Pesticides 73, 79
Pharaoh's ant 3
Phoretic nematodes 150
Physiological selectivity 25
Phytoseiid mite 9
Pink bollworm 43, 54, 74
Plant diversity 111
Polyphagous 5, 6
Polytrophic interaction 12
Polytrophic symbiosis 95
Potato aphid 43
Potato tuberworm 54
Praying mantids 5
Predacious mite 44, 48
Predators 1, 5, 54, 146, 155
Predatory mites 54
Preservation 164
Prickly pear cactus 165
Protozoans 146
Pteromalid parasitoid 24
Purple locoweed 164
Push Pull concept 2
Pyraloids 85
Pyrethroids 23, 73

Quality issues 67; control 71; environmental
    69; genetic 67

Rearing 53
Red locust 3
Red scale 54
Redox-based communication system 96
Release programs 8
Robber fly 5
Root and leaf feeding weevil 164
Rice stem borer 8
Rubbervine 167
Rubidium-marking technique 12

Salvinia 173
Salvinia weevil 165
San Jose scale 8

Scarab grubs 145
Semiochemical-mediated behavior 46
Silverleaf whitefly 31, 47
Skeleton weed 172
Soybean looper 91, 92
Spider mites 9, 23, 54
Spiders 5, 9
Spotted alfalfa aphid 22
Stem boring agromyzid 164
Stinkbugs 29
Stored grain pests 3
Sugarcane borers 8, 43, 44, 54, 71, 74, 128
Sugarcane pyrilla 8
Sugarcane shoot borer 8
Sugarcane top borer 8
Sulfur 79
Superparasitized 123
Sweet potato whitefly 23
Synergism 140, 142, 143, 146, 148; analysis
    142; host population dynamics 148;
    inteactions 144, 146, 147, 150, 151, 152,
    153; terminology 140
Synergistic 142
Synergistic effects 140
Syrphid fly 24, 151

Tachinid parasites 152
Tachinid parasitoid 146, 147, 150
Teflubenzuron 26
Temporal synergism 140, 147
Tephritid fruitfly 123
Teratocytes 152
Tobacco budworm 43, 44
Tomato fruit borer 8
Tomato fruit worm 44, 45
Tomato hornworm 44
Tortricid leafroller 45
Tritrophic interactions 11, 12, 87
Trophic levels 1, 2
Two-spotted spide mite 43, 48, 54

Vedalia beetle 3, 7
Vegetation diversity 111

Wasps 70, 71, 73, 76–78, 80
Water hyacinth 173, 176
Weeds 163; biocontrol procedures 166;
    classical biocontrol 165; evaluation 170;
    future of biocontrol 175
Western grape leafhopper 25, 48
Western grape skeletonizer 147
Whitefly 7, 47, 156
Whitefly control 47

Yellow striped armyworm 132